T0245264

CAMBRIDGE LIBRARY COLLECTION

Books of enduring scholarly value

Earth Sciences

In the nineteenth century, geology emerged as a distinct academic discipline. It pointed the way towards the theory of evolution, as scientists including Gideon Mantell, Adam Sedgwick, Charles Lyell and Roderick Murchison began to use the evidence of minerals, rock formations and fossils to demonstrate that the earth was older by millions of years than the conventional, Bible-based wisdom had supposed. They argued convincingly that the climate, flora and fauna of the distant past could be deduced from geological evidence. Volcanic activity, the formation of mountains, and the action of glaciers and rivers, tides and ocean currents also became better understood. This series includes landmark publications by pioneers of the modern earth sciences, who advanced the scientific understanding of our planet and the processes by which it is constantly re-shaped.

The Glaciers of the Alps

John Tyndall (1820–93) was an influential Irish geologist who became fascinated by mountaineering after a scientific expedition to Switzerland in 1856. He joined the Alpine Club in 1858 and achieved the summit of the Matterhorn in 1868 – a feat which led to a peak on the Italian side of the massif being named after him. He also climbed Mount Blanc three times. A writer of scientific texts who was widely praised for the quality of his prose, Tyndall made clear that in this work, published in 1860, he had 'not attempted to mix Narrative and Science'. He divides his account into two parts: his Alpine adventures and observations, and the scientific explanations about the origins and structural aspects of glaciers. Both sections include explanatory illustrations. This book, a classic text of Alpine exploration, offers a unique account of Tyndall's mountaineering expeditions and the science that inspired them.

Cambridge University Press has long been a pioneer in the reissuing of out-of-print titles from its own backlist, producing digital reprints of books that are still sought after by scholars and students but could not be reprinted economically using traditional technology. The Cambridge Library Collection extends this activity to a wider range of books which are still of importance to researchers and professionals, either for the source material they contain, or as landmarks in the history of their academic discipline.

Drawing from the world-renowned collections in the Cambridge University Library, and guided by the advice of experts in each subject area, Cambridge University Press is using state-of-the-art scanning machines in its own Printing House to capture the content of each book selected for inclusion. The files are processed to give a consistently clear, crisp image, and the books finished to the high quality standard for which the Press is recognised around the world. The latest print-on-demand technology ensures that the books will remain available indefinitely, and that orders for single or multiple copies can quickly be supplied.

The Cambridge Library Collection will bring back to life books of enduring scholarly value (including out-of-copyright works originally issued by other publishers) across a wide range of disciplines in the humanities and social sciences and in science and technology.

The Glaciers of the Alps

John Tyndall

CAMBRIDGE
UNIVERSITY PRESS

CAMBRIDGE UNIVERSITY PRESS

Cambridge, New York, Melbourne, Madrid, Cape Town,
Singapore, São Paolo, Delhi, Tokyo, Mexico City

Published in the United States of America by Cambridge University Press, New York

www.cambridge.org
Information on this title: www.cambridge.org/9781108037815

© in this compilation Cambridge University Press 2011

This edition first published 1860
This digitally printed version 2011

ISBN 978-1-108-03781-5 Paperback

THE MER DE GLACE.

Showing the Cleft Station at Trélaporte, les Echelets, the Tacul the Périades, and the Grande Jorasse.

THE

GLACIERS OF THE ALPS.

BEING

A NARRATIVE OF EXCURSIONS AND ASCENTS,

AN ACCOUNT OF THE ORIGIN AND PHENOMENA OF GLACIERS,

AND

AN EXPOSITION OF THE PHYSICAL PRINCIPLES TO WHICH THEY ARE RELATED.

By JOHN TYNDALL, F.R.S.,

MEMBER OF THE ROYAL SOCIETIES OF SCIENCE OF HOLLAND AND GÖTTINGEN; OF THE SCIENTIFIC
SOCIETIES OF HALLE, MARBURG, AND ZÜRICH; OF THE SOCIÉTÉ PHILOMATIQUE
OF PARIS; OF THE NATURAL HISTORY AND PHYSICAL SOCIETY OF
GENEVA; OF THE PHYSICAL SOCIETY OF BERLIN;
PROFESSOR OF NATURAL PHILOSOPHY IN THE ROYAL INSTITUTION OF GREAT BRITAIN,
AND IN THE GOVERNMENT SCHOOL OF MINES.

WITH ILLUSTRATIONS.

LONDON:
JOHN MURRAY, ALBEMARLE STREET.
1860.

The right of Translation is reserved.

TO

MICHAEL FARADAY,

THIS BOOK

IS AFFECTIONATELY INSCRIBED.

1860.

PREFACE.

In the following work I have not attempted to mix Narrative and Science, believing that the mind once interested in the one, cannot with satisfaction pass abruptly to the other. The book is therefore divided into Two Parts: the first chiefly narrative, and the second chiefly scientific.

In Part I. I have sought to convey some notion of the life of an Alpine explorer, and of the means by which his knowledge is acquired. In Part II. an attempt is made to classify such knowledge, and to refer the observed phenomena to their physical causes.

The Second Part of the work is written with a desire to interest intelligent persons who may not possess any special scientific culture. For their sakes I have dwelt more fully on principles than I should have done in presence of a purely scientific audience. The brief sketch of the nature of Light and Heat, with which Part II. is commenced, will not, I trust, prove uninteresting to the reader for whom it is more especially designed.

Should any obscurity exist as to the meaning of the terms Structure, Dirt-bands, Regelation, Interference, and others, which occur in Part I., it will entirely disappear in the perusal of Part II.

Two ascents of Mont Blanc and two of Monte Rosa are recorded; but the aspects of nature, and other circumstances which attracted my attention, were so different in the respective cases, that repetition was scarcely possible.

The numerous interesting articles on glaciers which have been published during the last eighteen months, and the various lively discussions to which the subject has given birth, have induced me to make myself better acquainted than I had previously been with the historic aspect of the question. In some important cases I have stated, with the utmost possible brevity, the results of my reading, and thus, I trust, contributed to the formation of a just estimate of men whose labours in this field were long anterior to my own.

J. T.

Royal Institution, June, 1860.

CONTENTS.

PART I.

1.—INTRODUCTORY.

2.—EXPEDITION OF 1856: THE OBERLAND.

3.—THE TYROL.

4.—EXPEDITION OF 1857: THE LAKE OF GENEVA.

5.—CHAMOUNI AND THE MONTANVERT.

b 3

24.

25.—SECOND ASCENT OF MONT BLANC, 1858.

26.

27.—WINTER EXPEDITION TO THE MER DE GLACE, 1859.

PART II.

1.—LIGHT AND HEAT.

2.—RADIANT HEAT.

3.—QUALITIES OF HEAT.

4.—ORIGIN OF GLACIERS.

5.

6.—Colour of Water and Ice.

7.—Colours of the Sky.

8.—The Moraines.

9.—Glacier Motion,—Preliminary.

10.—Motion of the Mer de Glace.

ILLUSTRATIONS.

PART I.

CHIEFLY NARRATIVE.

Ages are your days,
Ye grand expressors of the present tense
And types of permanence ;
Firm ensigns of the fatal Being
Amid these coward shapes of joy and grief
That will not bide the seeing.
Hither we bring
Our insect miseries to the rocks,
And the whole flight with pestering wing
Vanish and end their murmuring,
Vanish beside these dedicated blocks.

EMERSON.

GLACIERS OF THE ALPS.

INTRODUCTORY.

(1.)

In the autumn of 1854 I attended the meeting of the
British Association at Liverpool; and, after it was over,
availed myself of my position to make an excursion into
North Wales. Guided by a friend who knew the country,
I became acquainted with its chief beauties, and concluded
the expedition by a visit to Bangor and the neighbouring
slate quarries of Penrhyn.

From my boyhood I had been accustomed to handle
slates; had seen them used as roofing materials, and had
worked the usual amount of arithmetic upon them at
school; but now, as I saw the rocks blasted, the broken
masses removed to the sheds surrounding the quarry, and
there cloven into thin plates, a new interest was excited,
and I could not help asking after the cause of this extraordinary property of cleavage. It sufficed to strike the
point of an iron instrument into the edge of a plate of rock
to cause the mass to yield and open, as wood opens in
advance of a wedge driven into it. I walked round the
quarry and observed that the planes of cleavage were
everywhere parallel; the rock was capable of being split
in one direction only, and this direction remained perfectly
constant throughout the entire quarry.

I was puzzled, and, on expressing my perplexity to my
companion, he suggested that the cleavage was nothing

B

more than the layers in which the rock had been originally deposited, and which, by some subsequent disturbance,
had been set on end, like the strata of the sandstone rocks
and chalk cliffs of Alum Bay. But though I was too ignorant to combat this notion successfully, it by no means
satisfied me. I did not know that at the time of my visit
this very question of slaty cleavage was exciting the
greatest attention among English geologists, and I quitted
the place with that feeling of intellectual discontent which,
however unpleasant it may be for a time, is very useful as
a stimulant, and perhaps as necessary to the true appreciation of knowledge as a healthy appetite is to the enjoyment of food.

On inquiry I found that the subject had been treated
by three English writers, Professor Sedgwick, Mr. Daniel
Sharpe, and Mr. Sorby. From Professor Sedgwick I
learned that cleavage and stratification were things totally
distinct from each other; that in many cases the strata
could be observed with the cleavage passing through them
at a high angle; and that this was the case throughout vast
areas in North Wales and Cumberland. I read the lucid
and important memoir of this eminent geologist with great
interest: it placed the data of the problem before me, as
far as they were then known, and I found myself, to some
extent at least, in a, condition to appreciate the value of a
theoretic explanation.

Everybody has heard of the force of gravitation, and of
that of cohesion; but there is a more subtle play of forces
exerted by the molecules of bodies upon each other when
these molecules possess sufficient freedom of action. In
virtue of such forces, the ultimate particles of matter are
enabled to build themselves up into those wondrous edifices
which we call crystals. A diamond is a crystal self-erected
from atoms of carbon; an amethyst is a crystal built up
from particles of silica; Iceland spar is a crystal built

by particles of carbonate of lime. By artificial means we can allow the particles of bodies the free play necessary to their crystallization. Thus a solution of saltpetre exposed to slow evaporation produces crystals of saltpetre; alum crystals of great size and beauty may be obtained in a similar manner; and in the formation of a bit of common sugar-candy there are agencies at play, the contemplation of which, as mere objects of thought, is sufficient to make the wisest philosopher bow down in wonder, and confess himself a child.

The particles of certain crystalline bodies are found to arrange themselves in layers, like courses of atomic masonry, and along these layers such crystals may be easily cloven into the thinnest laminæ. Some crystals possess *one* such direction in which they may be cloven, some several; some, on the other hand, may be split with different facility in different directions. Rock salt may be cloven with equal facility in three directions at right angles to each other; that is, it may be split into cubes; calcspar may be cloven in three directions oblique to each other; that is, into rhomboids. Heavy spar may also be cloven in three directions, but one cleavage is much more perfect, or more *eminent* as it is sometimes called, than the rest. Mica is a crystal which cleaves very readily in one direction, and it is sufficiently tough to furnish films of extreme tenuity: finally, any boy, with sufficient skill, who tries a good crystal of sugar-candy in various directions with the blade of his penknife, will find that it possesses one direction in particular, along which, if the blade of the knife be placed and struck, the crystal will split into plates possessing clean and shining surfaces of cleavage.

Professor Sedgwick was intimately acquainted with all these facts, and a great many more, when he investigated the cleavage of slate rocks; and seeing no other explanation open to him, he ascribed to slaty cleavage a crystalline origin. He supposed that the particles of slate rock were acted on, after their deposition, by "polar forces," which so arranged them as to produce the cleavage. According to this theory, therefore, Honister Crag and the cliffs of Penrhyn are to be regarded as portions of enormous crystals; a length of time commensurate with the vastness of the supposed action being assumed to have elapsed between the deposition of the rock and its final crystallization.

When, however, we look closely into this bold and beautiful hypothesis, we find that the only analogy which exists between the physical structure of slate rocks and of crystals is this single one of cleavage. Such a coincidence might fairly give rise to the conjecture that both were due to a common cause; but there is great difficulty in accepting this as a theoretic truth. When we examine the structure of a slate rock, we find that the substance is composed of the débris of former rocks; that it was once a fine mud, composed of particles of *sensible magnitude*. Is it meant that these particles, each taken as a whole, were re-arranged after deposition? If so, the force which effected such an arrangement must be wholly different from that of crystallization, for the latter is essentially *molecular*. What is this force? Nature, as far as we know, furnishes none competent, under the conditions, to produce the effect. Is it meant that the molecules composing these sensible particles have re-arranged themselves? We find no evidence of such an action in the individual fragments: the mica is still mica, and possesses all the properties of mica; and so of the other ingredients of which the rock is composed. Independent of this, that an aggregate of heterogeneous mineral fragments should, without any assignable external cause, so shift its molecules as to produce a plane of cleavage common to them all, is, in my opinion, an assumption too heavy for any theory to bear.

Nevertheless, the paper of Professor Sedgwick invested the subject of slaty cleavage with an interest not to be forgotten, and proved the stimulus to further inquiry. The structure of slate-rocks was more closely examined; the fossils which they contained were subjected to rigid scrutiny, and their shapes compared with those of the same species taken from other rocks. Thus proceeding, the late Mr. Daniel Sharpe found that the fossils contained in slate-rocks are distorted in shape, being uniformly flattened out in the direction of the planes of cleavage. Here, then, was a fact of capital importance,—the shells became the indicators of an action to which the mass containing them had been subjected; they demonstrated the operation of pressure acting at right angles to the planes of cleavage.

The more the subject was investigated, the more clearly were the evidences of pressure made out. Subsequent to Mr. Sharpe, Mr. Sorby entered upon this field of inquiry. With great skill and patience he prepared sections of slate rock, which he submitted to microscopic examination, and his observations showed that the evidences of pressure could be plainly traced, even in his minute specimens. The subject has been since ably followed up by Professors Haughton, Harkness, and others; but to the two gentlemen first mentioned we are, I think, indebted for the prime facts on which rests the *mechanical theory* of slaty cleavage.*

The observations just referred to showed the co-existence of the two phenomena, but they did not prove that pressure and cleavage stood to each other in the relation of cause and effect. "Can the pressure produce the cleavage?" was still an open question, and it was one which mere reasoning, unaided by experiment, was incompetent to answer.

* Mr. Sorby has drawn my attention to an able and interesting paper by M. Bauer, in Karsten's 'Archiv' for 1846; in which it is announced that cleavage is a tension of the mass *produced by pressure*. The author refers to the experiments of Mr. Hopkins as bearing upon the question.

Sharpe despaired of an experimental solution, regarding our means as inadequate, and our time on earth too short to produce the result. Mr. Sorby was more hopeful. Submitting mixtures of gypsum and oxide of iron scales to pressure, he found that the scales set themselves approximately at right angles to the direction in which the pressure was applied. The position of the scales resembled that of the plates of mica which his researches had disclosed to him in slate-rock, and he inferred that the presence of such plates, and of flat or elongated fragments generally, lying all in the same general direction, was the cause of slaty cleavage. At the meeting of the British Association at Glasgow, in 1855, I had the pleasure of seeing some of Mr. Sorby's specimens, and, though the cleavage they exhibited was very rough, still, the tendency to yield at right angles to the direction in which the pressure had been applied, appeared sufficiently manifest.

At the time now referred to I was engaged, and had been for a long time previously, in examining the effects of pressure upon the magnetic force, and, as far back as 1851, I had noticed that some of the bodies which I had subjected to pressure exhibited a cleavage of surpassing beauty and delicacy. The bearing of such facts upon the present question now forcibly occurred to me. I followed up the observations; visited slate-yards and quarries, observed the exfoliation of rails, the fibres of iron, the structure of tiles, pottery, and cheese, and had several practical lessons in the manufacture of puff-paste and other laminated confectionery. My observations, I thought, pointed to a theory of slaty cleavage different from any previously given, and which, moreover, referred a great number of apparently unrelated phenomena to a common cause. On the 10th of June, 1856, I made them the subject of a Friday evening's discourse at the Royal Institution.

Such are the circumstances, apparently remote enough,

under which my connexion with glaciers originated. My friend, Professor Huxley, was present at the lecture referred to : he was well acquainted with the work of Professor Forbes, entitled 'Travels in the Alps,' and he surmised that the question of slaty cleavage, in its new aspect, might have some bearing upon the laminated structure of glacier-ice discussed in the work referred to. He therefore urged me to read the 'Travels,' which I did with care, and the book made the same impression upon me that it had produced upon my friend. We were both going to Switzerland that year, and it required but a slight modification of our plans to arrange a joint excursion over some of the glaciers of the Oberland, and thus afford ourselves the means of observing together the veined structure of the ice.

Had the results of this arrangement been revealed to me beforehand, I should have paused before entering upon an investigation which required of me so long a renunciation of my old and more favourite pursuits. But no man knows when he commences the examination of a physical problem into what new and complicated mental alliances it may lead him. No fragment of nature can be studied alone : each part is related to every other part; and hence it is that, following up the links of law which connect phenomena, the physical investigator often finds himself led far beyond the scope of his original intentions, the danger in this respect augmenting in direct proportion to the wish of the inquirer to render his knowledge solid and complete.

When the idea of writing this book first occurred to me, it was not my intention to confine myself to the glaciers alone, but to make the work a vehicle for the familiar explanation of such general physical phenomena as had come under my notice. Nor did I intend to address it to a cultured man of science, but to a *youth* of average intelligence, and furnished with the education which England now offers to the young. I wished indeed to make it a boy's class-book, which should reveal the mode of life, as well as the scientific objects, of an explorer of the Alps. The incidents of the past year have caused me to deviate, in some degree, from this intention, but its traces will be sufficiently manifest; and this reference to it will, I trust, excuse an occasional liberty of style and simplicity of treatment which would be out of place if intended for a reader of riper years.

EXPEDITION OF 1856.

THE OBERLAND.

(2.)

ON the 16th of August, 1856, I received my Alpenstock from the hands of Dr. Hooker, in the garden of the Pension Ober, at Interlaken. It bore my name, not marked, however, by the vulgar brands of the country, but by the solar beams which had been converged upon it by the pocket lens of my friend. I was the companion of Mr. Huxley, and our first aim was to cross the Wengern Alp. Light and shadow enriched the crags and green slopes as we advanced up the valley of Lauterbrunnen, and each occupied himself with that which most interested him. My companion examined the drift, I the cleavage, while both of us looked with interest at the contortions of the strata to our left, and at the shadowy, unsubstantial aspect of the pines, gleaming through the sunhaze to our right.

What was the physical condition of the rock when it was thus bent and folded like a pliant mass? Was it necessarily softer than it is at present? I do not think so. The shock which would crush a railway carriage, if communicated to it at once, is harmless when distributed over the interval necessary for the pushing in of the buffer. By suddenly stopping a cock from which water flows you may burst the conveyance pipe, while a slow turning of the cock keeps all safe. Might not a solid rock by ages of pressure be folded as above? It is a physical axiom that no body is perfectly hard, none perfectly soft, none perfectly elastic. The hardest body subjected to pressure yields, however little, and the same body when the pressure is removed *cannot return* to its original form. If it did not yield in the slightest degree it would be perfectly hard; if it could completely return to its original shape it would be perfectly elastic.

Let a pound weight be placed upon a cube of granite; the cube is flattened, though in an infinitesimal degree. Let the weight be removed, the cube *remains* a little flattened; it cannot quite return to its primitive condition. Let us call the cube thus flattened No. 1. Starting with No. 1 as a new mass, let the pound weight be laid upon it; the mass yields, and on removing the weight it cannot return to the dimensions of No. 1; we have a more flattened mass, No. 2. Proceeding in this manner, it is manifest that by a repetition of the process we should produce a series of masses, each succeeding one more flattened than the former. This appears to be a necessary consequence of the physical axiom referred to above.

Now if, instead of removing and replacing the weight in the manner supposed, we cause it to rest continuously upon the cube, the flattening, which above was intermittent, will be continuous; no matter how hard the cube may be, there will be a gradual yielding of its mass under the pressure. Apply this to squeezed rocks—to those, for example, which form the base of an obelisk like the Matterhorn; that this base must yield, seems a certain consequence of the physical constitution of matter: the conclusion seems inevitable that the mountain is sinking by its own weight. Let two points be fixed, one near the summit, the other near the base of the obelisk; next year these points will have approached each other. Whether the amount of approach in a human lifetime be measureable we know not; but it seems certain that ages would leave their impress upon the mass, and render visible to the eye an action which at present is appreciable by the imagination only.

We halted on the night of the 16th at the Jungfrau Hotel, and next morning we saw the beams of the rising sun fall upon the peaked snow of the Silberhorn. Slowly and solemnly the pure white cone appeared to rise higher and higher into the sunlight, being afterwards mottled with gold and gloom, as clouds drifted between it and the sun. I descended alone towards the base of the mountain, making my way through a rugged gorge, the sides of which were strewn with pine-trees, splintered, broken across, and torn up by the roots. I finally reached the end of a glacier, formed by the snow and shattered ice which fall from the shoulders of the Jungfrau. The view from this place had a savage magnificence such as I had not previously beheld, and it was not without some slight feeling of awe that I clambered up the end of the glacier. It was the first I had actually stood upon. The loneliness of the place was very impressive, the silence being only broken by fitful gusts of wind, or by the weird rattle of the débris which fell at intervals from the melting ice.

Once I noticed what appeared to be the sudden and enormous augmentation of the waters of a cascade, but the sound soon informed me that the increase was due to an avalanche which had chosen the track of the cascade for its rush. Soon afterwards my eyes were fixed upon a white slope some thousands of feet above me; I saw the ice give way, and, after a sensible interval, the thunder of another avalanche reached me. A kind of zigzag channel had been worn on the side of the mountain, and through this the avalanche rushed, hidden at intervals, and anon shooting forth, and leaping like a cataract down the precipices. The sound was sometimes continuous, but sometimes broken into rounded explosions which seemed to assert a passionate predominance over the general level of the roar. These avalanches, when they first give way, usually consist of enormous blocks of ice, which are more and more shattered as they descend. Partly to the echos of the first crash, but mainly, I think, to the shock of the harder masses which preserve their cohesion, the explosions which occur during the descent of the avalanche are to be ascribed. Much of the ice is crushed to powder; and thus, when an avalanche pours cataract-like over a ledge, the heavier masses, being less influenced by the atmospheric resistance, shoot forward like descending rockets, leaving the lighter powder in trains behind them. Such is the material of which a class of the smaller glaciers in the Alps is composed. They are the products of avalanches, the crushed ice being recompacted into a solid mass, which exhibits on a smaller scale most of the characteristics of the large glaciers.

After three hours' absence I reascended to the hotel, breakfasted, and afterwards returned with Mr. Huxley to the glacier. While we were engaged upon it the weather suddenly changed; lightning flashed about the summits of the Jungfrau, and thunder "leaped" among her crags. Heavy rain fell, but it cleared up afterwards with magical speed, and we returned to our hotel. Heedless of the forebodings of many prophets of evil weather we set out for Grindelwald. The scene from the summit of the little Scheideck was exceedingly grand. The upper air exhibited a commotion which we did not experience; clouds were wildly driven against the flanks of the Eiger, the Jungfrau thundered behind, while in front of us a magnificent rainbow, fixing one of its arms in the valley of Grindelwald, and, throwing the other right over the crown of the Wetterhorn, clasped the mountain in its embrace. Through jagged apertures in the clouds floods of golden light were poured down the sides of the mountain. On the slopes were innumerable châlets, glistening in the sunbeams, herds browsing peacefully and shaking their mellow bells; while the blackness of the pine-trees, crowded

into woods, or scattered in pleasant clusters over alp and
valley, contrasted forcibly with the lively green of the
fields.

At Grindelwald, on the 18th, we engaged a strong and
competent guide, named Christian Kaufmann, and proceeded to the Lower Glacier. After a steep ascent, we
gained a point from which we could look down upon the
frozen mass. At first the ice presented an appearance of
utter confusion, but we soon reached a position where the
mechanical conditions of the glacier revealed themselves,
and where we might learn, had we not known it before,
that confusion is merely the unknown intermixture of laws,
and becomes order and beauty when we rise to their comprehension. We reached the so-called Eismeer—Ice Sea.
In front of us was the range of the Viescherhörner, and a
vast snow slope, from which one branch of the glacier
was fed. Near the base of this *névé*, and surrounded on
all sides by ice, lay a brown rock, to which our attention
was directed as a place noted for avalanches; on this
rock snow or ice never rests, and it is hence called the
Heisse Platte—the Hot Plate. At the base of the rock,
and far below it, the glacier was covered with clean crushed
ice, which had fallen from a crown of frozen cliffs en-
circling the brow of the rock. One obelisk in particular
signalised itself from all others by its exceeding grace and
beauty. Its general surface was dazzling white, but from
its clefts and fissures issued a delicate blue light, which
deepened in hue from the edges inwards. It stood upon a
pedestal of its own substance, and seemed as accurately
fixed as if rule and plummet had been employed in its
erection. Fig. 1 represents this beautiful minaret of ice.

While we were in sight of the *Heisse Platte*, a dozen
avalanches rushed downwards from its summit. In most
cases we were informed of the descent of an avalanche by
the sound, but sometimes the white mass was seen gliding

down the rock, and scattering its *smoke* in the air, long
before the sound reached us. It is difficult to reconcile the
insignificant appearance presented by avalanches, when seen
from a distance, with the volume of sound which they

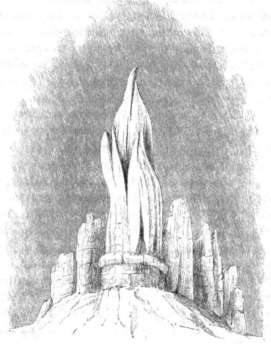

Fig. 1.

generate; but on this day we saw sufficient to account for
the noise. One block of solid ice which we found below
the *Heisse Platte* measured 7 feet 6 inches in length, 5 feet
8 inches in height, and 4 feet 6 inches in depth. A second
mass was 10 feet long, 8 feet high, and 6 feet wide. It
contained therefore 480 cubic feet of ice, which had been
cast to a distance of nearly 1,000 yards down the glacier.
The shock of such hard and ponderous projectiles against
rocks and ice, reinforced by the echos from the surrounding

mountains, will appear sufficient to account for the peals by which their descent is accompanied.

A second day, in company with Dr. Hooker, completed the examination of this glacier in 1856; after which I parted from my friends, Mr. Huxley intending to rejoin me at the Grimsel. On the morning of the 20th of August I strapped on my knapsack and ascended the green slopes from Grindelwald towards the great Scheideck. Before reaching the summit I frequently heard the wonderful echos of the Wetterhorn. Some travellers were in advance of me, and to amuse *them* an alpine horn was blown. The direct sound was cut off from me by a hill, but the echos talked down to me from the mountain walls. The sonorous waves arrived after one, two, three, and more reflections, diminishing gradually in intensity, but increasing in softness, as if in its wanderings from crag to crag the sound had undergone a kind of sifting process, leaving all its grossness behind, and returning in delightful flute notes to the ear.

Let us investigate this point a little. If two looking-glasses be placed perfectly parallel to each other, with a lighted candle between them, an infinite series of images of the candle will be seen at both sides, the images diminishing in brightness the further they recede. But if the looking-glasses, instead of being parallel, enclose an angle, a limited number of images only will be seen. The smaller the angle which the reflectors make with each other, or, in other words, the nearer they approach parallelism, the greater will be the number of images observed.

To find the number of images the following is the rule :— Divide 360, or the number of degrees in a circle, by the number of degrees in the angle enclosed by the two mirrors, the quotient will be *one more* than the number of images; or, counting the object itself, the quotient is always equal to the number of images plus the object. In Fig. 2 I have

given the number and position of the images produced by
two mirrors placed at an angle of 45°. A B and B C mark

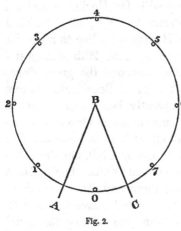
Fig. 2.

the edges of the mirrors,
and O represents the can-
dle, which, for the sake
of simplicity, I have
placed midway between
them. Fix one point of
a pair of compasses at
B, and with the distance
B O sweep a circle:—*all
the images will be ranged
upon the circumference of
this circle.* The number
of images found by the
foregoing rule is 7, and
their positions are marked in the figure by the numbers 1,
2, 3, &c.

Suppose the *ear* to occupy the place of the eye, and
that *a sounding body* occupies the place of the luminous
one, we should then have just as many *echos* as we had
images in the former case. These echos would diminish
in loudness just as the images of the candle diminish in
brightness. At each reflection a portion both of sound and
light is lost; hence the oftener light is reflected the dimmer
it becomes, and the oftener sound is reflected the fainter
it is.

Now the cliffs of the Wetterhorn are so many rough
angular reflectors of the sound : some of them send it back
directly to the listener, and we have a first echo; some of
them send it on to others from which it is again reflected, forming a second echo. Thus, by repeated reflection, successive echos are sent to the ear, until, at length,
they become so faint as to be inaudible. The sound, as
it diminishes in intensity, appears to come from greater

and greater distances, as if it were receding into the mountain solitudes; the final echos being inexpressibly soft and
pure.

After crossing the Scheideck I descended to Meyringen,
visiting the Reichenbach waterfall on my way. A pecu-
liarity of the descending water here is, that it is broken up
in one of the basins into nodular masses, each of which in
falling leaves the light foaming mass which surrounds it as
a train in the air behind; the effect exactly resembles
that of the avalanches of the Jungfrau, in which the more
solid blocks of ice shoot forward in advance of the lighter
débris, which is held back by the friction of the air.

Next day I ascended the valley of Hasli, and observed
upon the rocks and mountains the action of ancient glaciers
which once filled the valley to the height of more than a
thousand feet above its present level. I paused, of course,
at the waterfall of Handeck, and stood for a time upon
the wooden bridge which spans the river at its top. The
Aar comes gambolling down to the bridge from its parent
glacier, takes one short jump upon a projecting ledge,
boils up into foam, and then leaps into a chasm, from the
bottom of which its roar ascends through the gloom. A
rivulet named the Aarlenbach joins the Aar from the left
in the very jaws of the chasm: falling, at first, upon a
projection at some depth below the edge, and, rebound-
ing from this, it darts at the Aar, and both plunge
together like a pair of fighting demons to the bottom of
the gorge. The foam of the Aarlenbach is white, that
of the Aar is yellow, and this enables the observer to trace
the passage of the one cataract *through* the other. As I
stood upon the bridge the sun shone brightly upon the
spray and foam; my shadow was oblique to the river, and
hence a symmetrical rainbow could not be formed in the
spray, but one half of a lovely bow, with its base in the
chasm, leaned over against the opposite rocks, the colours

advancing and retreating as the spray shifted its position. I had been watching the water intently for some time, when a little Swiss boy, who stood beside me, observed, in his trenchant German, " There plunge stones ever downwards." The stones were palpable enough, carried down by the cataract, and sometimes completely breaking loose from it, but I did not see them until my attention was withdrawn from the water.

On my arrival at the Grimsel I found Mr. Huxley already there, and, after a few minutes' conversation, we decided to spend a night in a hut built by M. Dolfuss in 1846, beside the Unteraar glacier, about 2000 feet above the Hospice. We hoped thus to be able to examine the glacier to its origin on the following day. Two days' food and some blankets were sent up from the Hospice, and, accompanied by our guide, we proceeded to the glacier.

Having climbed a great terminal moraine, and tramped for a considerable time amid loose shingle and boulders, we came upon the ice. The finest specimens of "tables" which I have ever seen are to be found upon this glacier —huge masses of clean granite poised on pedestals of ice. Here are also "dirt-cones" of the largest size, and nume-rous shafts, the forsaken passages of ancient "moulins," some filled with water, others simply with deep blue light. I reserve the description and explanation of both cones and moulins for another place. The surfaces of some of the small pools were sprinkled lightly over with snow, which the water underneath was unable to melt; a coating of snow granules was thus formed, flexible as chain armour, but so close that the air could not escape through it. Some bubbles which had risen through the water had lifted the coating here and there into little rounded domes, which, by gentle pressure, could be shifted hither and thither, and several of them collected into one. We reached the hut, the floor of which appeared to be of the original moun-

tain slab; there was a space for cooking walled off from
the sleeping-room, half of which was raised above the floor,
and contained a quantity of old hay. The number 2404
mètres, the height, I suppose, of the place above the sea,
was painted on the door, behind which were also the names
of several well-known observers—Agassiz, Forbes, Desor,
Dolfuss, Ramsay, and others—cut in the wood. A loft contained a number of instruments for boring, a surveyor's
chain, ropes, and other matters. After dinner I made my
way alone towards the junction of the Finsteraar and
Lauteraar glaciers, which unite at the Abschwung to form
the trunk stream of the Unteraar glacier. Upon the great
central moraine which runs between the branches were
perched enormous masses of rock, and, under the overhanging ledge of one of these, M. Agassiz had his *Hôtel
des Neufchâtelois*. The rock is still there, bearing traces of
names now nearly obliterated by the weather, while the
fragments around also bear inscriptions. There in the
wilderness, in the gray light of evening, these blurred and
faded evidences of human activity wore an aspect of sad-
ness. It was a temple of science now in ruins, and I a
solitary pilgrim to the desecrated blocks. As the day declined, rain began to fall, and I turned my face towards
my new home; where in due time we betook ourselves to
our hay, and waited hopefully for the morning.

But our hopes were doomed to disappointment. A vast
quantity of snow fell during the night, and, when we arose,
we found the glacier covered, and the air thick with the
descending flakes. We waited, hoping that it might clear
up, but noon arrived and passed without improvement;
our fire-wood was exhausted, the weather intensely cold,
and, according to the men's opinion, hopelessly bad;
they opposed the idea of ascending further, and we had
therefore no alternative but to pack up and move downwards. What was snow at the higher elevations changed

to rain lower down, and drenched us completely before we reached the Grimsel. But though thus partially foiled in our design, this visit taught us much regarding the structure and general phenomena of the glacier.

The morning of the 24th was clear and calm: we rose with the sun, refreshed and strong, and crossed the Grimsel pass at an early hour. The view from the summit of the pass was lovely in the extreme; the sky a deep blue, the surrounding summits all enamelled with the newly-fallen snow, which gleamed with dazzling whiteness in the sunlight. It was Sunday, and the scene was itself a Sabbath, with no sound to disturb its perfect rest. In a lake which we passed the mountains were mirrored without distortion, for there was no motion of the air to ruffle its surface. From the summit of the Mayenwand we looked down upon the Rhone glacier, and a noble object it seemed,—I hardly know a finer of its kind in the Alps. Forcing itself through the narrow gorge which holds the ice cascade in its jaws, and where it is greatly riven and dislocated, it spreads out in the valley below in such a manner as clearly to reveal to the mind's eye the nature of the forces to which it is subjected. Longfellow's figure is quite correct; the glacier resembles a vast gauntlet, of which the gorge represents the wrist; while the lower glacier, cleft by its fissures into finger-like ridges, is typified by the hand.

Furnishing ourselves with provisions at the adjacent inn, we devoted some hours to the examination of the lower portion of the glacier. The dirt upon its surface was arranged in grooves as fine as if produced by the passage of a rake, while the laminated structure of the deeper ice always corresponded to the superficial grooving. We found several shafts, some empty, some filled with water. At one place our attention was attracted by a singular noise, evidently produced by the forcing of air and water through passages in the body of the glacier; the sound rose and fell for several minutes, like a kind of intermittent snore, reminding one of Hugi's hypothesis that the glacier was alive.

We afterwards climbed to a point from which the whole glacier was visible to us from its origin to its end. Adjacent to us rose the mighty mass of the Finsteraarhorn, the monarch of the Oberland. The Galenstock was also at hand, while round about the *névé* of the glacier a mountain wall projected its jagged outline against the sky. At a distance was the grand cone of the Weisshorn, then, and I believe still, unscaled; further to the left the magnificent peaks of the Mischabel; while between them, in savage isolation, stood the obelisk of the Matterhorn. Near us was the chain of the Furca, all covered with shining snow, while overhead the dark blue of the firmament so influenced the general scene as to inspire a sentiment of wonder approaching to awe. We descended to the glacier, and proceeded towards its source. As we advanced an unusual light fell upon the mountains, and looking upwards we saw a series of coloured rings, drawn like a vivid circular rainbow quite round the sun. Between the orb and us spread a thin veil of cloud on which the circles were painted; the western side of the veil soon melted away, and with it the colours, but the eastern half remained a quarter of an hour longer, and then in its turn disappeared. The crevasses became more frequent and dangerous as we ascended. They were usually furnished with overhanging eaves of snow, from which long icicles depended, and to tread on which might be fatal. We were near the source of the glacier, but the time necessary to reach it was nevertheless indefinite, so great was the entanglement of fissures. We followed one huge chasm for some hundreds of yards, hoping to cross it; but after half an hour's fruitless effort we found ourselves baffled and forced to retrace our steps.

The sun was sloping to the west, and we thought it wise
to return; so down the glacier we went, mingling our
footsteps with the tracks of chamois, while the frightened
marmots piped incessantly from the rocks. We reached
the land once more, and halted for a time to look upon
the scene within view. The marvellous blueness of the
sky in the earlier part of the day indicated that the air was
charged, almost to saturation, with transparent aqueous
vapour. As the sun sank the shadow of the Finsteraarhorn was cast through the adjacent atmosphere, which,
thus deprived of the direct rays, curdled up into visible
fog. The condensed vapour moved slowly along the flanks
of the mountain, and poured itself cataract-like into the
valley of the Rhone. Here it met the sun again, which
reduced it once more to the invisible state. Thus, though
there was an incessant supply from the generator behind,
the fog made no progress; as in the case of the moving
glacier, the end of the cloud-river remained stationary
where consumption was equal to supply. Proceeding along
the mountain to the Furca, we found the valley at the
further side of the pass also filled with fog, which rose, like
a wall, high above the region of actual shadow. Once on
turning a corner an exclamation of surprise burst simultaneously from my companion and myself. Before each
of us and against the wall of fog, stood a spectral image
of a man, of colossal dimensions; dark as a whole, but
bounded by a coloured outline. We stretched forth our
arms; the spectres did the same. We raised our alpen-
stocks; the spectres also flourished their batons. All our
actions were imitated by these fringed and gigantic shades.
We had, in fact, *the Spirit of the Brocken* before us in perfection.

At the time here referred to I had had but little experience of alpine phenomena. I had been through the
Oberland in 1850, but was then too ignorant to learn

much from my excursion. Hence the novelty of this day's experience may have rendered it impressive : still even now I think there was an intrinsic grandeur in its phenomena which entitles the day to rank with the most remarkable that I have spent among the Alps. At the Furca, to my great regret, the joint ramblings of my friend and myself ended ; I parted from him on the mountain side, and watched him descending, till the gray of evening finally hid him from my view.

THE TYROL.

(3.)

My subsequent destination was Vienna ; but I wished to associate with my journey thither a visit to some of the glaciers of the Tyrol. At Landeck, on the 29th of August, I learned that the nearest glacier was that adjacent to the Gebatsch Alp, at the head of the Kaunserthal ; and on the following morning I was on my way towards this valley. I sought to obtain a guide at Kaltebrunnen, but failed ; and afterwards walked to the little hamlet of Feuchten, where I put up at a very lowly inn. My host, I believe, had never seen an Englishman, but he had heard of such, and remarked to me in his patois with emphasis, "Die Engländer sind die kühnste Leute in dieser Welt." Through his mediation I secured a chamois-hunter, named Johann Auer, to be my guide, and next morning I started with this man up the valley. The sun, as we ascended, smote the earth and us with great power ; high mountains flanked us on either side, while in front of us, closing the view, was the mass of the Weisskugel, covered with snow. At three o'clock we came in sight of the glacier, and soon afterwards I made the acquaintance of the *Senner* or cheesemakers of the Gebatsch Alp.

The chief of these was a fine tall fellow, with free, frank countenance, which, however, had a dash of the mountain wildness in it. His feet were bare, he wore breeches, and fragments of stockings partially covered his legs, leaving a black zone between the upper rim of the sock and the breeches. His feet and face were of the same swarthy hue; still he was handsome, and in a measure pleasant to look upon. He asked me what he could cook for me, and I requested some bread and milk; the former was a month old, the latter was fresh and delicious, and on these I fared sumptuously. I went to the glacier afterwards with my guide, and remained upon the ice until twilight, when we returned, guided by no path, but passing amid crags grasped by the gnarled roots of the pine, through green dells, and over bilberry knolls of exquisite colouring. My guide kept in advance of me singing a Tyrolese melody, and his song and the surrounding scene revived and realised all the impressions of my boyhood regarding the Tyrol.

Milking was over when we returned to the châlet, which now contained four men exclusive of myself and my guide. A fire of pine logs was made upon a platform of stone, elevated three feet above the floor; there was no chimney, as the smoke found ample vent through the holes and fissures in the sides and roof. The men were all intensely sunburnt, the legitimate brown deepening into black with beard and dirt. The chief senner prepared supper, breaking eggs into a dish, and using his black fingers to empty the shell when the albumen was refractory. A fine erect figure he was as he stood in the glowing light of the fire. All the men were smoking, and now and then a brand was taken from the fire to light a renewed pipe, and

a ruddy glare flung thereby over the wild countenance of
the smoker. In one corner of the châlet, and raised high
above the ground, was a large bed, covered with clothes
of the most dubious black-brown hue; at one end was a
little water-wheel turned by a brook, which communicated motion to a churndash which made the butter. The
beams and rafters were covered with cheeses, drying in the
warm smoke. The senner, at my request, showed me his
storeroom, and explained to me the process of making
cheese, its interest to me consisting in its bearing upon
the question of slaty cleavage. Three gigantic masses of
butter were in the room, and I amused my host by calling
them butter-glaciers. Soon afterwards a bit of cotton was
stuck in a lump of grease, which was placed in a lantern,
and the wick ignited; the chamois-hunter took it, and led
the way to our resting-place, I having previously declined
a good-natured invitation to sleep in the big black bed
already referred to.

There was a cowhouse near the châlet, and above it,
raised on pillars of pine, and approached by a ladder, was
a loft, which contained a quantity of dry hay: this my
guide shook to soften the lumps, and erected an eminence
for my head. I lay down, drawing my plaid over me, but
Auer affirmed that this would not be a sufficient protec-
tion against the cold; he therefore piled hay upon me to
the shoulders, and proposed covering up my head also.
This, however, I declined, though the biting coldness of
the air, which sometimes blew in upon us, afterwards
proved to me the wisdom of the suggestion. Having set
me right, my chamois-hunter prepared a place for himself,
and soon his heavy breathing informed me that he was
in a state of bliss which I could only envy. One by one
the stars crossed the apertures in the roof. Once the
Pleiades hung above me like a cluster of gems; I tried to
admire them, but there was no fervour in my admiration.

Sometimes I dozed, but always as this was about to deepen into positive sleep it was rudely broken by the clamour of a group of pigs which occupied the ground-floor of our dwelling. The object of each individual of the group was to secure for himself the maximum amount of heat, and hence the outside members were incessantly trying to become inside ones. It was the struggle of radical and conservative among the pachyderms, the politics being determined by the accident of position.

I rose at five o'clock on the 1st of September, and after a breakfast of black bread and milk ascended the glacier as far as practicable. We once quitted it, crossed a promontory, and descended upon one of its branches, which was flanked by some fine old moraines. We here came upon a group of seven marmots, which with yells of terror scattered themselves among the rocks. The points of the glacier beyond my reach I examined through a telescope; along the faces of the sections the lines of stratification were clearly shown; and in many places where the mass showed manifest signs of lateral pressure, I thought I could observe the cleavage passing through the strata. The point, however, was too important to rest upon an observation made from such a distance, and I therefore abstained from mentioning it subsequently. I examined the fissures and the veining, and noticed how the latter became most perfect in places where the pressure was greatest. The effect of *oblique* pressure was also finely shown: at one place the thrust of the descending glacier was opposed by the resistance offered by the side of the valley, the direction of the force being oblique to the side; the consequence was a structure nearly parallel to the valley, and consequently oblique to the thrust which I believe to be its cause.

After five hours' examination we returned to our châlet, where we refreshed ourselves, put our things in order, and

faced a nameless "Joch," or pass; our aim being to cross
the mountains into the valley of Lantaufer, and reach
Graun that evening. After a rough ascent over the alp
we came to the dead crag, where the weather had broken
up the mountains into ruinous heaps of rock and shingle.
We reached the end of a glacier, the ice of which was
covered by sloppy snow, and at some distance up it came
upon an islet of stones and débris, where we paused to rest
ourselves. My guide, as usual, ranged over the summits
with his telescope, and at length exclaimed, " I see a
chamois." The creature stood upon a cliff some hundreds
of yards to our left, and seemed to watch our movements.
It was a most graceful animal, and its life and beauty stood
out in forcible antithesis to the surrounding savagery and
death.

On the steep slopes of the glacier I was assisted by the
hand of my guide. In fact, on this day I deemed places
dangerous, and dreaded them as such, which subsequent
practice enabled me to regard with perfect indifference;
so much does what we call courage depend upon habit, or
on the fact of knowing that we have really nothing to fear.
Doubtless there are times when a climber has to make up
his mind for very unpleasant possibilities, and even gather
calmness from the contemplation of the worst; but in most
cases I should say that his courage is derived from the
latent feeling that the chances of safety are immensely in
his favour.

After a tough struggle we reached the narrow row of
crags which form the crest of the pass, and looked into
the world of mountain and cloud on the other side. The
scene was one of stern grandeur—the misty lights and
deep cloud-glooms being so disposed as to augment the
impression of vastness which the scene conveyed. The
breeze at the summit was exceedingly keen, but it gave
our muscles tone, and we sprang swiftly downward

through the yielding débris which here overlies the mountain, and in which we sometimes sank to the knees.
Lower down we came once more upon the ice. The
glacier had at one place melted away from its bounding
cliff, which rose vertically to our right, while a wall of ice
60 or 80 feet high was on our left. Between both was
a narrow passage, the floor of which was snow, which
I knew to be hollow beneath: my companion, however,
was in advance of me, and he being the heavier man,
where he trod I followed without hesitation. On turning an angle of the rock I noticed an expression of
concern upon his countenance, and he muttered audibly,
"I did not expect this." The snow-floor had, in fact,
given way, and exposed to view a clear green lake, one
boundary of which was a sheer precipice of rock, and the
other the aforesaid wall of ice; the latter, however, curved
a little at its base, so as to form a short steep slope
which overhung the water. My guide first tried the slope
alone; biting the ice with his shoe-nails, and holding on
by the spike of his baton, he reached the other side. He
then returned, and, divesting myself of all superfluous
clothes, as a preparation for the plunge which I fully
expected, I also passed in safety. Probably the consciousness that I had water to fall into instead of pure space,
enabled me to get across without anxiety or mischance;
but had I, like my guide, been unable to swim, my feelings
would have been far different.

This accomplished, we went swiftly down the valley,
and the more I saw of my guide the more I liked him.
He might, if he wished, have made his day's journey
shorter by stopping before he reached Graun, but he
would not do so. Every word he said to me regarding
distances was true, and there was not the slightest desire
shown to magnify his own labour. I learnt by mere accident that the day's work had cut up his feet, but his

cheerfulness and energy did not bate a jot till he had landed
me in the Black Eagle at Graun. Next morning he came
to my room, and said that he felt sufficiently refreshed
to return home. I paid him what I owed him, when
he took my hand, and, silently bending down his head,
kissed it; then, standing erect, he stretched forth his
right hand, which I grasped firmly in mine, and bade him
farewell; and thus I parted from Johann Auer, my brave
and truthful chamois-hunter.

On the following day I met Dr. Frankland in the Finster-
muntz pass, and that night we bivouacked together at
Mals. Heavy rain fell throughout the night, but it came
from a region high above that of liquidity. It was first
snow, which, as it descended through the warmer strata
of the atmosphere, was reduced to water. Overhead, in
the air, might be traced a surface, below which the precipitate was liquid, above which it was solid; and this surface, intersecting the mountains which surround Mals,
marked upon them a beautifully-defined *snow-line*, below
which the pines were dark and the pastures green, but
above which pines and pastures and crags were covered
with the freshly-fallen snow.

On the 2nd of September we crossed the Stelvio. The
brown cone of the well-known Madatschspitze was clear, but
the higher summits were clouded, and the fragments of sunshine which reached the lower world wandered like gleams
of fluorescent light over the glaciers. Near the snow-line
the partial melting of the snow had rendered it coarsely
granular, but as we ascended it became finer, and the light
emitted from its cracks and cavities a pure and deep
blue. When a staff was driven into the snow low down
the mountain, the colour of the light in the orifice was
scarcely sensibly blue, but higher up this increased in a
wonderful degree, and at the summit the effect was marvellous. I struck my staff into the snow, and turned it

round and round; the surrounding snow cracked repeat-
edly, and flashes of blue light issued from the fissures.
The fragments of snow that adhered to the staff were, by
contrast, of a beautiful pink yellow, so that, on moving the
staff with such fragments attached to it up and down, it
was difficult to resist the impression that a pink flame was
ascending and descending in the hole. As we went down
the other side of the pass, the effect became more and
more feeble, until, near the snow-line, it almost wholly disappeared.

  We remained that night at the baths of Bormio, but, the
following afternoon being fine, we wished to avail ourselves
of the fair weather to witness the scene from the summit of
the pass. Twilight came on before we reached Santa Maria,
but a gorgeous orange overspread the western horizon, from
which we hoped to derive sufficient light. It was a little
too late when we reached the top, but still the scene was
magnificent. A multitude of mountains raised their crowns
towards heaven, while above all rose the snow-white cone
of the Ortler. Far into the valley the giant stretched his
granite limbs, until they were hid from us by darkness. As
this deepened, the heavens became more and more crowded
with stars, which blazed like gems over the heads of the
mountains. At times the silence was perfect, unbroken
save by the crackling of the frozen snow beneath our own
feet; while at other times a breeze would swoop down
upon us, keen and hostile, scattering the snow from the roofs
of the wooden galleries in frozen powder over us. Long
after night had set in, a ghastly gleam rested upon the
summit of the Ortler, while the peaks in front deepened to
a dusky neutral tint, the more distant ones being lost in
gloom. We descended at a swift pace to Trafoi, which we
reached before 11 P.M.

  Meran was our next resting-place, whence we turned
through the Schnalzerthal to Unserfrau, and thence over

the Hochjoch to Fend. From a religious procession we
took a guide, who, though partly intoxicated, did his duty
well. Before reaching the summit of the pass we were
assailed by a violent hailstorm, each hailstone being a
frozen cone with a rounded end. Had not their motion
through the air something to do with the shape of these
hailstones? The theory of meteorites now generally
accepted is that they are small planetary bodies drawn to
the earth by gravity, and brought to incandescence by
friction against the earth's atmosphere. Such a body
moving through the atmosphere must have condensed hot
air in front of it, and rarefied cool air behind it; and the
same is true to a small extent of a hailstone. This distribution of temperature must, I imagine, have some influence
on the shape of the stone. Possibly also the stratified appearance of some hailstones may be connected with this
action.*

The hail ceased and the heights above us cleared as we
ascended. At the top of the pass we found ourselves on
the verge of a great *névé*, which lay between two ranges of
summits, sloping down to the base of each range from a
high and rounded centre: a wilder glacier scene I have
scarcely witnessed. Wishing to obtain a more perfect
view of the region, I diverged from the track followed by
Dr. Frankland and the guide, and climbed a ridge of snow
about half a mile to the right of them. A glorious expanse

* I take the following account of a grander storm of the above character from Hooker's 'Himalayan Journals,' vol. ii. p. 405.

"On the 20th (March, 1849) we had a change in the weather: a violent
storm from the south-west occurred at noon, with hail of a strange form,
the stones being sections of hollow spheres, half an inch across and upwards, formed of cones with truncated apices and convex bases: these
cones were aggregated together with their bases outwards. The large
masses were followed by a shower of the separate conical pieces, and that
by heavy rain On the mountains this storm was most severe: the stones
lay at Dorjiling for seven days, congealed into masses of ice several feet
long and a foot thick in sheltered places: at Purneah, fifty miles south,
stones one and two inches across fell, probably as whole spheres."

was before me, stretching itself in vast undulations, and heaping itself here and there into mountainous cones, white and pure, with the deep blue heaven behind them. Here I had my first experience of hidden crevasses, and to my extreme astonishment once found myself in the jaws of a fissure of whose existence I had not the slightest notice. Such accidents have often occurred to me since, but the impression made by the first is likely to remain the strongest. It was dark when we reached the wretched Wirthshaus at Fend, where, badly fed, badly lodged, and disturbed by the noise of innumerable rats, we spent the night. Thus ended my brief glacier expedition of 1856; and on the observations then made, and on subsequent experiments, was founded a paper presented to the Royal Society by Mr. Huxley and myself.

EXPEDITION OF 1857.

THE LAKE OF GENEVA.

(4.)

THE time occupied in the observations of 1856 embraced about five whole days; and though these days were laborious and instructive, still so short a time proved to be wholly incommensurate with the claims of so wide a problem. During the subsequent experimental treatment of the subject, I had often occasion to feel the incompleteness of my knowledge, and hence arose the desire to make a second expedition to the Alps, for the purpose of expanding, fortifying, or, if necessary, correcting first impressions.

On Thursday, the 9th of July, 1857, I found myself upon the Lake of Geneva, proceeding towards Vevey. I had long wished to see the waters of this renowned inland sea, the colour of which is perhaps more interesting to the man of science than to the poets who have sung about it. Long ago its depth of blue excited attention, but no systematic examination of the subject has, so far as I know, been attempted. It may be that the lake simply exhibits the colour of pure water. Ice is blue, and it is reasonable to suppose that the liquid obtained from the fusion of ice is of the same colour; but still the question presses—"Is the blue of the Lake of Geneva to be entirely accounted for in this way?" The attempts which have been made to explain it otherwise show that at least a doubt exists as to the sufficiency of the above explanation.

It is only in its deeper portions that the colour of the lake is properly seen. Where the bottom comes into view the pure effect of the water is disturbed; but where the water is deep the colour is deep. between Rolle and Nyon for example, the blue is superb. Where the blue was deepest, however, it gave me the impression of turbidity rather than of deep transparency. At the upper portion of the lake the water through which the steamer passed was of a blue green. Wishing to see the place where the Rhone enters the lake, I walked on the morning of the 10th from Villeneuve to Novelle, and thence through the woods to the river side. Proceeding along an embankment, raised to defend the adjacent land from the incursions of the river, an hour brought me to the place where it empties itself into the lake. The contrast between the two waters was very great: the river was almost white with the finely divided matter which it held in suspension; while the lake at some distance was of a deep ultramarine.

The lake in fact forms a reservoir where the particles held in suspension by the river have time to subside, and its waters to become pure. The subsidence of course takes place most copiously at the head of the lake; and here the deposit continues to form new land, adding year by year to the thousands of acres which it has already left behind it, and invading more and more the space occupied by the water. Innumerable plates of mica spangled the fine sand which the river brought down, and these, mixing with the water, and flashing like minute mirrors as the sun's rays fell upon them, gave the otherwise muddy stream a silvery appearance. Had I an opportunity I would make the following experiments:—

(*a.*) Compare the colour of the light transmitted by a column of the lake water fifteen feet long with that transmitted by a second column, of the same length, derived from the melting of freshly fallen mountain snow.

(*b.*) Compare in the same manner the colour of the ordinary water of the lake with that of the same water after careful distillation.

(*c.*) Strictly examine whether the light transmitted by the ordinary water contains an excess of red over that transmitted by the distilled water: this latter point, as will be seen farther on, is one of peculiar interest.

The length is fixed at fifteen feet, because I have found this length extremely efficient in similar experiments.

On returning to the pier at Villeneuve, a peculiar flickering motion was manifest upon the surface of the distant portions of the lake, and I soon noticed that the coast line was inverted by atmospheric refraction. It required a long distance to produce the effect: no trace of it was seen about the Castle of Chillon, but at Vevey and beyond it, the whole coast was clearly inverted; and the houses on the margin of the lake were also imaged to a certain height. Two boats at a considerable distance presented the appearance sketched in Figs. 3 and 4; the hull

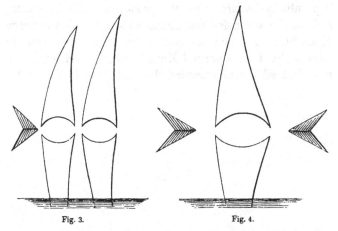

Fig. 3.   Fig. 4.

of each, except a small portion at the end, was invisible, but the sails seemed lifted up high in the air, with their inverted images below; as the boats drew nearer the hulls appeared inverted, the apparent height of the vessel above the surface of the lake being thereby nearly doubled, while the sails and higher objects, in these cases, were almost completely cut away. When viewed through a telescope the sensible horizon of the lake presented a billowy tumultuous appearance, fragments being incessantly detached from it and suspended in the air.

The explanation of this effect is the same as that of the mirage of the desert, which may be found in almost any book on physics, and which so tantalized the French soldiers in Egypt. They often mistook this aërial inversion for the reflection from a lake, and on trial found hot and sterile sand at the place where they expected refreshing waters. The effect was shown by Monge, one of the learned men who accompanied the expedition, to be due to the total reflection of very oblique rays at the upper surface of the layer of rarefied air which was nearest to the heated earth. A sandy plain, in the early part of the day, is peculiarly favourable for the production of such effects; and on the extensive flat strand which stretches between Mont St. Michel and the coast adjacent to Avranches in Normandy, I have noticed Mont Tombeline reflected as if glass instead of sand surrounded it and formed its mirror.

CHAMOUNI AND THE MONTANVERT.

(5.)

On the evening of the 12th of July I reached Chamouni; the weather was not quite clear, but it was promising; white cumuli had floated round Mont Blanc during the day, but these diminished more and more, and the light of the setting sun was of that lingering rosy hue which bodes good weather.   Two parallel beams of a purple tinge were drawn by the shadows of the adjacent peaks, straight across the Glacier des Bossons, and the Glacier des Pélerins was also steeped for a time in the same purple light.   Once when the surrounding red illumination was strong, the shadows of the Grands Mulets falling upon the adjacent snow appeared of a *vivid green.*

This green belonged to the class of *subjective* colours, or colours produced by contrast, about which a volume might be written.   The eye received the impression of green, but the colour was not external to the eye.   Place a red wafer on white paper, and look at it intently, it will be surrounded in a little time by a green fringe : move the wafer bodily away, and the entire space which it occupied upon the paper will appear green.   A body may have its proper colour entirely masked in this way.   Let a red wafer be attached to a piece of red glass, and from a moderately illuminated position let the sky be regarded through the glass; the wafer will appear of a vivid green. If a strong beam of light be sent through a red glass and caused to fall upon a screen, which at the same time is moderately illuminated by a separate source of white light, an opaque body placed in the path of the beam will cast a green shadow upon the screen which may be seen by several hundred persons at once. If a blue glass be used, the shadow will be yellow, which is the complementary colour to blue.

When we suddenly pass from open sunlight to a moderately illuminated room, it appears dark at first, but after a little time the eye regains the power of seeing objects distinctly. Thus one effect of light upon the eye is to render it less sensitive, and light of any particular colour falling upon the eye blunts its appreciation *of that colour*. Let us apply this to the shadow upon the screen. This shadow is moderately illuminated by a jet of white light; but the space surrounding it is red, the effect of which upon the eye is to *blind* it in some degree to the perception of red. Hence, when the feeble white light of the shadow reaches the eye, the red component of this light is, as it were, abstracted from it, and the eye sees the residual colour, which is green. A similar explanation applies to the shadows of the Grands Mulets.

On the 13th of July I was joined by my friend Mr. Thomas Hirst, and on the 14th we examined together the end of the Mer de Glace. In former times the whole volume of the Arveiron escaped from beneath the ice at the end of the glacier, forming a fine arch at its place of issue. This year a fraction only of the water thus found egress; the greater portion of it escaping laterally from the glacier at the summit of the rocks called *Les Motets*, down which it tumbled in a fine cascade. The vault at the end of the glacier was nevertheless respectable, and rather tempting to a traveller in search of information regarding the structure of the ice. Perhaps, however, Nature meant to give me a friendly warning at the outset. for, while speculating as to the wisdom of entering the cavern, it suddenly gave way, and, with a crash which rivalled thunder, the roof strewed itself in ruins upon the floor.

Many years ago I had read with delight Coleridge's

poem entitled 'Sunrise in the Valley of Chamouni,' and
to witness in all perfection the scene described by the poet,
I waited at Chamouni a day longer than was otherwise
necessary. On the morning of Wednesday, the 15th of July,
I rose before the sun; Mont Blanc and his wondrous staff of
Aiguilles were without a cloud; eastward the sky was of a
pale orange which gradually shaded off to a kind of rosy
violet, and this again blended by imperceptible degrees
with the deep zenithal blue. The morning star was still
shining to the right, and the moon also turned a pale face
towards the rising day. The valley was full of music;
from the adjacent woods issued a gush of song, while the
sound of the Arve formed a suitable bass to the shriller melody of the birds. The mountain rose for a time cold and
grand, with no apparent stain upon his snows. Suddenly the
sunbeams struck his crown and converted it into a boss of
gold. For some time it remained the only gilded summit
in view, holding communion with the dawn while all the
others waited in silence. These, in the order of their
heights, came afterwards, relaxing, as the sunbeams struck
each in succession, into a blush and smile.

On the same day we had our luggage transported to the
Montanvert, while we clambered along the lateral moraine
of the glacier to the Chapeau. The rocks alongside the
glacier were beautifully scratched and polished, and I paid
particular attention to them, for the purpose of furnishing
myself with a key to ancient glacier action. The scene
to my right was one of the most wonderful I had ever witnessed. Along the entire slope of the Glacier de Bois, the
ice was cleft and riven into the most striking and fantastic
forms. It had not yet suffered much from the wasting influence of the summer weather, but its towers and minarets
sprang from the general mass with clean chiselled outlines.
Some stood erect, others leaned, while the white *débris*,
strewn here and there over the glacier, showed where the

wintry edifices had fallen, breaking themselves to pieces,
and grinding the masses on which they fell to powder.
Some of them gave way during our inspection of the place,
and shook the valley with the reverberated noise of their
fall. I endeavoured to get near them, but failed; the
chasms at the margin of the glacier were too dangerous, and
the stones resting upon the heights too loosely poised to
render persistence in the attempt excusable.

We subsequently crossed the glacier to the Montanvert,
and I formally took up my position there. The rooms of
the hotel were separated from each other by wooden par-
titions merely, and thus the noise of early risers in one
room was plainly heard in the next. For the sake of
quiet, therefore, I had my bed placed in the *château* next
door,—a little octagonal building erected by some
kind and sentimental Frenchman, and dedicated " *à la
Nature.*" My host at first demurred, thinking the place
not "*propre,*" but I insisted, and he acquiesced. True the
stone floor was dark with moisture, and on the walls a
glistening was here and there observable, which suggested
rheumatism, and other penalties, but I had had no experience of rheumatism, and trusted to the strength which
mountain air and exercise were sure to give me, for
power to resist its attacks. Moreover, to dispel some of
the humidity, it was agreed that a large pine fire should
be made there on necessary occasions.

Though singularly favoured on the whole, still our resi-
dence at the Montanvert was sufficiently long to give us
specimens of all kinds of weather; and thus my château
derived an interest from the mutations of external nature.
Sometimes no breath disturbed the perfect serenity of the
night, and the moon, set in a black-blue sky, turned a face
of almost supernatural brightness to the mountains, while
in her absence the thick-strewn stars alone flashed and
twinkled through the transparent air. Sometimes dull

dank fog choked the valley, and heavy rain plashed upon the stones outside.   On two or three occasions we were favoured by a thunderstorm, every peal of which broke into a hundred echos, while the seams of lightning which ran through the heavens produced a wonderful intermittence of gloom and glare.   And as I sat within, musing on the experiences of the day, with my pine logs crackling, and the ruddy fire-light gleaming over the walls, and lending animation to the visages sketched upon them with charcoal by the guides, I felt that my position was in every way worthy of a student of nature.

THE MER DE GLACE.

(6.)

The name "Mer de Glace" has doubtless led many who have never seen this glacier to a totally erroneous conception of its character. Misled probably by this term, a distinguished writer, for example, defines a glacier to be a sheet of ice spread out upon the slope of a mountain; whereas the Mer de Glace is indeed a *river*, and not a *sea* of ice. But certain forms upon its surface, often noticed and described, and which I saw for the first time from the window of our hotel on the morning of the 16th of July, suggest at once the origin of the name. The glacier here has the appearance of a sea which, after it had been tossed by a storm, had suddenly stiffened into rest. The ridges upon its surface accurately resemble waves in shape, and this singular appearance is produced in the following way :—

Some distance above the Montanvert—opposite to the Echelets—the glacier, in passing down an incline, is rent by deep fissures, between each two of which a ridge of ice intervenes. At first the edges of these ridges are sharp and angular, but they are soon sculptured off by the action of the sun. The bearing of the Mer de Glace being approximately north and south, the sun at mid-day shines down the glacier, or rather very obliquely across it; and the consequence is, that the fronts of the ridges, which look downward, remain in shadow all the day, while the backs of the ridges, which look up the glacier, meet the direct stroke of the solar rays. The ridges thus acted upon have their hindmost angles wasted off and converted

into slopes which represent the *back* of a wave, while the
opposite sides of the ridges, which are protected from the
sun, preserve their steepness, and represent the *front* of the
wave.   Fig. 5 will render my meaning at once plain.

Fig. 5.

The dotted lines are intended to represent three of the
ridges into which the glacier is divided, with their interposed fissures; the dots representing the boundaries of the
ridges when the glacier is first broken.   The parallel
shading lines represent the direction of the sun's rays,
which, falling obliquely upon the ridges, waste away the
right-hand corners, and finally produce wave-like forms.

We spent a day or two in making the general acquaintance of the glacier.   On the 16th we ascended till we came
to the rim of the Talèfre basin, from which we had a good
view of the glacier system of the region.   The laminated structure of the ice was a point which particularly
interested me; and as I saw the exposed sections of the
névé, counted the lines of stratification, and compared
these with the lines upon the ends of the secondary gla-

ciers, I felt the absolute necessity either of connecting
the veined *structure* with the *strata* by a continuous chain
of observations, or of proving by ocular evidence that
they were totally distinct from each other. I was well
acquainted with the literature of the subject, but nothing
that I had read was sufficient to prove what I required.
Strictly speaking, nothing that had been written upon the
subject rose above the domain of *opinion*, while I felt that
without absolute *demonstration* the question would never be
set at rest.

On this day we saw some fine glacier tables; flat masses
of rock, raised high upon columns of ice: Fig. 6 is a

Fig. 6.

sketch of one of the finest of them. Some of them fell
from their pedestals while we were near them, and the
clean ice-surfaces which they left behind sparkled with
minute stars as the small bubbles of air ruptured the film
of water by which they were overspread. I also noticed that
" petit bruit de crépitation," to which M. Agassiz alludes,
and which he refers to the rupture of the ice by the ex-

pansion of the air-bubbles contained within it.   When I
first read Agassiz's account of it, I thought it might be
produced by the rupture of the minute air-bubbles which
incessantly escape from the glacier.   This, doubtless, produces an effect, but there is something in the character of
the sound to be referred, I think, to a less obvious cause,
which I shall notice further on.

At six p.m. this day I reached the Montanvert; and the
same evening, wrapping my plaid around me, I wandered
up towards Charmoz, and from its heights observed, as
they had been observed fifteen years previously by Professor Forbes, the *dirt-bands* of the Mer de Glace.   They
were different from any I had previously seen, and I felt
a strong desire to trace them to their origin.   Content,
however, with the performance of the day, and feeling
healthily tired by it, I lay down upon the bilberry bushes
and fell asleep.   It was dark when I awoke, and I experienced some difficulty and risk in getting down from the
petty eminence referred to.

The illumination of the glacier, as remarked by Professor Forbes, has great influence upon the appearance of
the bands; they are best seen in a subdued light, and I
think for the following reasons :—

The dirt-bands are seen simply because they send less
light to the eye than the cleaner portions of the glacier
which lie between them; two surfaces, differently illuminated, are presented to the eye, and it is found that this
difference is more observable when the light is that of
evening than when it is that of noon.

It is only within certain limits that the eye is able to
perceive differences of intensity in different lights; beyond
a certain intensity, if I may use the expression, light ceases
to be light, and becomes mere pain.   The naked eye can
detect no difference in brightness between the electric light

and the lime light, although, when we come to strict measurement, the former may possess many times the intensity of the latter. It follows from this that we might reduce the ordinary electric light to a fraction of its intensity, without any perceptible change of brightness to the naked eye which looks at it. But if we reduce the lime light in the same proportion the effect would be very different. This light lies much nearer to the limit at which the eye can appreciate differences of brightness, and its reduction might bring it quite within this limit, and make it sensibly dimmer than before. Hence we see that when two sources of intense light are presented to the eye, by reducing both the lights in the same proportion, the *difference* between them may become more perceptible.

Now the dirt-bands and the spaces between them resemble, in some measure, the two lights above mentioned. By the full glare of noon both are so strongly illuminated that the difference which the eye perceives is very small; as the evening advances the light of both is lowered in the same proportion, but the *differential* effect upon the eye is thereby augmented, and the bands are consequently more clearly seen.

(7.)

On Friday, the 17th of July, we commenced our measurements. Through the kindness of Sir Roderick Murchison, I found myself in the possession of an excellent five-inch theodolite, an instrument with the use of which both my friend Hirst and myself were perfectly familiar.

We worked in concert for a few days to familiarize our assistant with the mode of proceeding, but afterwards it was my custom to simply determine the position where a measurement was to be made, and to leave the execution of it entirely to Mr. Hirst and our guide.

On the 20th of July I made a long excursion up the glacier, examining the moraines, the crevasses, the structure, the moulins, and the disintegration of the surface. I was accompanied by a boy named Edouard Balmat,* and found him so good an iceman that I was induced to take him with me on the following day also.

Looking upwards from the Montanvert to the left of the Aiguille de Charmoz, a singular gap is observed in the rocky mountain wall, in the centre of which stands a detached column of granite. Both cleft and pillar are shown in the frontispiece, to the right. The eminence to the left of this gap is signalised by Professor Forbes as one of the best stations from which to view the Mer de Glace, and this point, which I shall refer to hereafter as the *Cleft Station*, it was now my desire to attain. From the Montanvert side a steep gully leads to the cleft; up this couloir we proposed to try the ascent. At a considerable height above the Mer de Glace, and closely hugging the base of the Aiguille de Charmoz, is the small Glacier de Tendue, shown in the frontispiece, and from which a steep slope stretches down to the Mer de Glace. This Tendue is the most *talkative* glacier I have ever known; the clatter of the small stones which fall from it is incessant. Huge masses of granite also frequently fall upon the glacier from the cliffs above it, and, being slowly borne downwards by the moving ice, are at length seen toppling above the terminal face of the glacier. The ice which supports them being gradually melted, they are

* "Le petit Balmat" my host always called him.

at length undermined, and sent bounding down the slope
with peal and rattle, according as the masses among which
they move are large or small. The space beneath the
glacier is cumbered with blocks thus sent down; some of
them of enormous size.

The danger arising from this intermittent cannonade,
though in reality small, has caused the guides to swerve
from the path which formerly led across the slope to the
promontory of Trélaporte. I say "small," because, even
should a rock choose the precise moment at which a traveller is passing to leap down, the boulders at hand are so
large and so capable of bearing a shock that the least
presence of mind would be sufficient to place him in safety.
But presence of mind is not to be calculated on under
such circumstances, and hence the guides were right to
abandon the path.

Reaching the mouth of our gully after a rough ascent,
we took to the snow, instead of climbing the adjacent
rocks. It was moist and soft, in fact in a condition altogether favourable for the "regelation" of its granules.
As the foot pressed upon it the particles became cemented
together. A portion of the pressure was transmitted
laterally, which produced attachments beyond the boundary of the foot; thus as the latter sank, it pressed upon
a surface which became continually wider and more rigid,
and at length sufficiently strong to bear the entire
weight of the body; the pressed snow formed in fact a
virtual *camel's foot*, which soon placed a limit to the sinking. It is this same principle of regelation which enables
men to cross snow bridges in safety. By gentle cautious
pressure the loose granules of the substance are cemented
into a continuous mass, all sudden shocks which might
cause the frozen surfaces to snap asunder being avoided.
In this way an arch of snow fifteen or twenty inches in

thickness may be rendered so firm that a man will cross it, although it may span a chasm one hundred feet in depth.

As we ascended, the incline became very steep, and once or twice we diverged from the snow to the adjacent rocks; these were disintegrated, and the slightest disturbance was sufficient to bring them down; some fell, and from one of them I found it a little difficult to escape; for it grazed my leg, inflicting a slight wound as it passed. Just before reaching the cleft at which we aimed, the snow for a short distance was exceedingly steep, but we surmounted it; and I sat for a time beside the granite pillar, pleased to find that I could permit my legs to dangle over a precipice without prejudice to my head.

While we remained here a chamois made its appearance upon the rocks above us. Deeming itself too near, it climbed higher, and then turned round to watch us. It was soon joined by a second, and both formed a very pretty picture: their attitudes frequently changed, but they were always graceful; with head erect and horns curved back, a light limb thrown forward upon a ledge of rock, looking towards us with wild and earnest gaze, each seemed a type of freedom and agility. Turning now to the left, we attacked the granite tower, from which we purposed to scan the glacier, and were soon upon its top. My companion was greatly pleased —he was "très-content" to have reached the place—he felt assured that many old guides would have retreated from that ugly gully, with its shifting shingle and débris, and his elation reached its climax in the declaration that, if I resolved to ascend Mont Blanc without a guide, he was willing to accompany me.

From the position which we had attained, the prospect was exceedingly fine, both of the glaciers and of the mountains. Beside us was the Aiguille de Charmoz, piercing with its spikes of granite the clear air. To my mind it is one of the finest of the Aiguilles, noble in mass, with its summits singularly cleft and splintered. In some atmospheric colourings it has the exact appearance of a mountain of cast copper, and the manner in which some of its highest pinnacles are bent, suggesting the idea of ductility, gives strength to the illusion that the mass is metallic. At the opposite side of the glacier was the Aiguille Verte, with a cloud poised upon its point: it has long been the ambition of climbers to scale this peak, and on this day it was attempted by a young French count with a long retinue of guides. He had not fair play, for before we quitted our position we heard the rumble of thunder upon the mountain, which indicated the presence of a foe more terrible than the avalanches themselves. Higher to the right, and also at the opposite side of the glacier, rose the Aiguille du Moine; and beyond was the basin of the Talèfre, the ice cascade issuing from which appeared, from our position, like the foam of a waterfall. Then came the Aiguille de Léchaud, the Petite Jorasse, the Grande Jorasse, and the Mont Tacul; all of which form a cradle for the Glacier de Léchaud. Mont Mallet, the Périades, and the Aiguille Noire, came next, and then the singular obelisk of the Aiguille du Géant, from which a serrated edge of cliff descends to the summit of the "Col."

Over the slopes of the Col du Géant was spread a coverlet of shining snow, at some places apparently as smooth as polished marble, at others broken so as to form precipices, on the pale blue faces of which the horizontal lines of bedding were beautifully drawn. As the eye approaches the line which stretches from the Rognon to the Aiguille Noire, the repose of the *névé* becomes more and more disturbed. Vast chasms are formed, which however are still merely indicative of the trouble in advance. If the glacier were lifted off we should probably see that the line just referred to would lie along the summit of a steep gorge; over this summit the glacier is pushed, and has its back periodically broken, thus forming vast transverse ridges which follow each other in succession down the slope. At the summit these ridges are often cleft by fissures transverse to *them,* thus forming detached towers of ice of the most picturesque and imposing character.* These towers often fall; and while some are caught upon the platforms of the cascade, others struggle with the slow energy of a behemoth through the débris which opposes them, reach the edges of the precipices which rise in succession along the fall, leap over, and, amid ice-smoke and thunder-peals, fight their way downwards.

A great number of secondary glaciers were in sight hanging on the steep slopes of the mountains, and from them streams sped downwards, falling over the rocks, and filling the valley with a low rich music. In front of me, for example, was the Glacier du Moine, and I could not help feeling as I looked at it, that the arguments drawn from the deportment of such glaciers against the " sliding theory," and which are still repeated in works upon the Alps, militate just as strongly against the "viscous theory." " How," demands the antagonist of the sliding theory, " can a secondary glacier exist upon so steep a slope? why does it not slide down as an avalanche?" " But how," the person addressed may retort, " can a mass which you assume to be viscous exist under similar conditions? If it be viscous, what prevents it from rolling down?" The sliding theory assumes the lubrication of the bed of the glacier, but on this cold height the quantity melted is too small

---

* To such towers the name *Sérac* is applied. In the châlets of Savoy, after the richer curd has been precipitated by rennet, a stronger acid is used to throw down what remains; an inferior kind of cheese called *Sérac* is thus formed, the shape and colour of which have suggested the application of the term to the cubical masses of ice.

to lubricate the bed, and hence the slow motion of these glaciers. Thus a sliding-theory man might reason, and, if the external deportment of secondary glaciers were to decide the question, De Saussure might perhaps have the best of the argument.

And with regard to the current idea, originated by M. de Charpentier, and adopted by Professor Forbes, that if a glacier slides it must slide as an avalanche, it may be simply retorted that, in part, *it does so ;* but if it be asserted that an *accelerated motion* is the necessary motion of an avalanche, the statement needs qualification. An avalanche on passing through a rough couloir soon attains an uniform velocity—its motion being accelerated only up to the point when the sum of the resistances acting upon it is equal to the force drawing it downwards. These resistances are furnished by the numberless asperities which the mass encounters, and which incessantly check its descent, and render an accumulation of motion impossible. The motion of a man walking down stairs may be on the whole uniform, but it is really made up of an aggregate of small motions, each of which is accelerated; and it is easy to conceive how a glacier moving over an uneven bed, when released from one opposing obstacle will be checked by another, and its motion thus rendered sensibly uniform.

From the Aiguille du Géant and Les Périades a glacier descended, which was separated by the promontory of La Noire from the glacier proceeding from the Col du Géant. A small moraine was formed between them, which is marked *a* upon the diagram, Fig. 7. The great mass of the glacier descending from the Col du Géant came next, and this was bounded on the side nearest to Trélaporte by a small moraine *b*, the origin of which I could not see, its upper portion being shut out by a mountain promontory. Between the moraine *b* and the actual side of the valley was another little glacier, derived from some of the lateral tributaries.

It was, however, between the moraines *a* and *b* that the
great mass of the Glacier du Géant really lay. At the

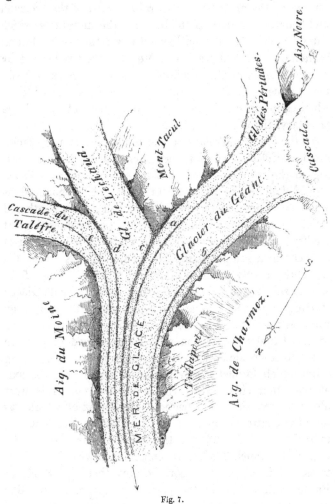

Fig. 7.

promontory of the Tacul the lateral moraines of the glacier des Périades and of the Glacier de Léchaud united to

form the medial moraine *c* of the Mer de Glace. Carrying the eye across the Léchaud, we had the moraine *d* formed by the union of the lateral moraines of the Léchaud and Talèfre ; further to the left was the moraine *e*, which came from the Jardin, and beyond it was the second lateral moraine of the Talèfre. The Mer de Glace is formed by the confluence of the whole of the glaciers here named; being forced at Trélaporte through a passage, the width of which appears considerably less than that of the single tributary, the Glacier du Géant.

In the ice near Trélaporte the blue veins of the glacier are beautifully shown ; but they vary in distinctness according to the manner in which they are looked at. When regarded obliquely their colour is not so pronounced as when the vision plunges deeply into them. The weathered ice of the surface near Trélaporte could be cloven with great facility ; I could with ease obtain plates of it a quarter of an inch thick, and possessing two square feet of surface. On the 28th of July I followed the veins several times from side to side across the Géant portion of the Mer de Glace ; starting from one side, and walking along the veins, my route was directed obliquely downwards towards the axis of the tributary. At the axis I was forced to turn, in order to keep along the veins, and now ascended along a line which formed nearly the same angle with the axis at the other side. Thus the veins led me as it were along the two sides of a triangle, the vertex of which was near the centre of the glacier. The vertex was, however, in reality rounded off, and the figure rather resembled a hyperbola, which tended to coincidence with its asymptotes. This observation corroborates those of Professor Forbes with regard to the position of the veins, and, like him, I found that at the centre the veining, whose normal direction would be transverse to the glacier, was contorted and confused.

Near the side of the Glacier du Géant, above the promontory of Trélaporte, the ice is rent in a remarkable
manner. Looking upwards from the lower portions of the
glacier, a series of vertical walls, rising apparently one
above the other, face the observer. I clambered up among
these singular terraces, and now recognise, both from my
sketch and memory, that their peculiar forms are due to
the same action as that which has given their shape to the
"billows" of the Mer de Glace. A series of profound
crevasses is first formed. The Glacier du Géant deviates
14° from the meridian line, and hence the sun shines
nearly down it during the middle portion of each day.
The backs of the ridges between the crevasses are thus
rounded off, one boundary of each fissure is destroyed, or
at least becomes a mere steep declivity, while the other
boundary being shaded from the sun preserves its ver-
ticality; and thus a very curious series of precipices is
formed.

Through all this dislocation, the little moraine on
which I have placed the letter b in the sketch maintains
its right to existence, and under it the laminated structure of this portion of the glacier appears to reach its
most perfect development. The moraine was generally
a mere dirt track, but one or two immense blocks of
granite were perched upon it. I examined the ice underneath one of these, being desirous of seeing whether the
pressure resulting from its enormous weight would produce a veining, but the result was not satisfactory.
Veins were certainly to be seen in directions different
from the normal ones, but whether they were due to
the bending of the latter, or were directly owing to the
pressure of the block, I could not say. The sides of a
stream which had cut a deep gorge in the clean ice
of the Glacier du Géant afforded a fine opportunity of
observing the structure. It was very remarkable—highly

significant indeed in a theoretic point of view. Two long
and remarkably deep blue veins traversed the bottom of
the stream, and bending upwards at a place where the
rivulet curved, drew themselves like a pair of parallel lines
upon the clean white ice. But the general structure was
of a totally different character; it did not consist of long
bars, but approximated to the lenticular form, and was,
moreover, of a washy paleness, which scarcely exceeded in
depth of colouring the whitish ice around.

To the investigator of the structure nothing can be finer
than the appearance of the glacier from one of the ice
terraces cut in the Glacier du Géant by its passage round
Trélaporte. As far as the vision extended the dirt upon
the surface of the ice was arranged in striæ. These striæ
were not always straight lines, nor were they unbroken
curves. Within slight limits the various parts into which
a glacier is cut up by its crevasses enjoy a kind of independent motion. The grooves, for example, on two ridges
which have been separated by a small fissure, may one
day have their striæ perfect continuations of each other,
but in a short time this identity of direction may be
destroyed by a difference of motion between the ridges.
Thus it is that the grooves upon the surface above Trélaporte are bent hither and thither, a crack or seam always
marking the point where their continuity is ruptured.
This bending occurs, however, within limits sufficiently
small to enable the striæ to preserve the same general
direction.

My attention had often been attracted this day by projecting masses of what at first appeared to be pure white
snow, rising in seams above the general surface of the
glacier. On examination, however, I found them to be
compact ice, filled with innumerable air-cells, and so
resistant as to maintain itself in some places at a height of
four feet above the general level. When amongst the

ridges they appeared discontinuous and confused, being
scattered apparently at random over the glacier; but when
viewed from a sufficient distance, the detached parts
showed themselves to belong to a system of white seams
which swept quite across the Glacier du Géant, in a direction concentric with the structure. Unable to account for
these singular seams, I climbed up among the tributary
glaciers on the Rognon side of the Glacier du Géant, and
remained there until the sun sank behind the neighbouring peaks, and the fading light warned me that it was
time to return.

(8.)

Early on the following day I was again upon the ice.
I first confined myself to the right side of the Glacier du
Géant, and found that the veins of white ice which I had
noticed on the previous day were exclusively confined to
this glacier, or to the space between the moraines a and
b (Fig. 7), bending up so that the moraine a between
the Glacier du Géant and the Glacier des Périades was
tangent to them. At a good distance up the glacier I
encountered a considerable stream rushing across it
almost from side to side. I followed the rivulet, examining the sections which it exposed. At a certain point
three other streams united, and formed at their place
of confluence a small green lake. From this a rivulet
rushed, which was joined by the stream whose track I had
pursued, and at this place of junction a second green lake
was formed, from which flowed a stream equal in volume
to the sum of all the tributaries. It entered a crevasse,
and took the bottom of the fissure for its bed. Standing
at the entrance of the chasm, a low muffled thunder

resounding through the valley attracted my attention. I
followed the crevasse, which deepened and narrowed, and,
by the blue light of the ice, could see the stream gambolling along its bottom, and flashing as it jumped over
the ledges which it encountered in its way. The fissure
at length came to an end: placing a foot on each side
of it, and withholding the stronger light from my eyes,
I looked down between its shining walls, and saw the
stream plunge into a shaft which carried it to the bottom
of the glacier.

Slowly, and in zigzag fashion, as the crevasses demanded,
I continued to ascend, sometimes climbing vast humps of
ice from which good views of the surrounding glacier
were obtained; sometimes hidden in the hollows between
the humps, in which also green glacier tarns were often
formed, very lonely and very beautiful.

While standing beside one of these, and watching the
moving clouds which it faithfully mirrored, I heard the
sound of what appeared to be a descending avalanche, but
the time of its continuance surprised me. Looking through
my opera-glass in the direction of the sound, I saw issuing
from the end of a secondary glacier on the Tacul side a torrent of what appeared to me to be stones and mud. I could
see the stones and finer débris jumping down the declivi-
ties, and shaping themselves into singular cascades. The
noise continued for a quarter of an hour, after which the
torrent rapidly diminished, until, at length, the ordinary
little stream due to the melting of the glacier alone
remained. A subglacial lake had burst its boundary, and
carried along with it in its rush downwards the débris
which it met with in its course.

In some places I found the crevasses difficult, the ice
being split in a very singular manner. Vast plates of
it not more than a foot in thickness were sometimes detached from the sides of the crevasses, and stood alone.

I was now approaching the base of the *séracs*, and the glacier around me still retained a portion of the turbulence of the cascade. I halted at times amid the ruin and confusion, and examined with my glass the cascade itself. It was a wild and wonderful scene, suggesting throes of spasmodic energy, though, in reality, all its dislocation had been *slowly* and *gradually* produced. True, the stratified blocks which here and there cumbered the terraces suggested *débacles*, but these were local and partial, and did not affect the general question. There is scarcely a case of geological disturbance which could not be matched with its analogue upon the glaciers,—contortions, faults, fissures, joints, and dislocations,—but in the case of the ice we can prove the effects to be due to slowly-acting causes; how reasonable is it then to ascribe to the operation of similar causes, which have had an incomparably longer time to work, many geological effects which at first sight might suggest sudden convulsion !

Wandering slowly upwards, successive points of attraction drawing me almost unconsciously on, I found myself as the day was declining deep in the entanglements of the ice. A shower commenced, and a splendid rainbow threw an oblique arch across the glacier. I was quite alone; the scene was exceedingly impressive, and the possibility of difficulties on which I had not calculated intervening between me and the lower glacier, gave a tinge of anxiety to my position. I turned towards home; crossed some bosses of ice and rounded others; I followed the tracks of streams which were very irregular on this portion of the glacier, bending hither and thither, rushing through deep-cut channels, falling in cascades and expanding here and there to deep green lakes; they often plunged into the depths of the ice, flowed under it with hollow gurgle, and reappeared at some distant point. I threaded my way cautiously amid systems of crevasses, scattering with my axe, to secure a footing, the rotten ice of the sharper crests, which fell with a ringing sound into the chasms at either side. Strange subglacial noises were sometimes heard, as if caverns existed underneath, into which blocks of ice fell at intervals, transmitting the shock of their fall with a dull boom to the surface of the glacier. By the steady surmounting of difficulties one after another, I at length placed them all behind me, and afterwards hastened swiftly along the glacier to my mountain home.

On the 30th incessant rain confined us to indoor work; on the 31st we determined the velocity with which the glacier is forced through the entrance of the trunk valley at Trélaporte, and also the motion of the Grand Moulin. We also determined both the velocity and the width of the Glacier du Géant. The 1st of August was spent by me at the cascade of the Talèfre, examining the structure, crumpling, and scaling off of the ice. Finding that the rules at Chamouni put an unpleasant limit to my demands on my guide Simond, I visited the Guide Chef on the 2nd of August, and explained to him the object of my expedition, pointing out the inconvenience which a rigid application of the rules made for tourists would impose upon me. He had then the good sense to acknowledge the reasonableness of my remarks, and to grant me the liberty I requested. The 3rd of August was employed in determining the velocity and width of the Glacier de Léchaud, and in observations on the lamination of the glacier.

THE JARDIN.

(9.)

ON the 4th of August, with a view of commencing a
series of observations on the inclinations of the Mer de
Glace and its tributaries, we had our theodolite transported
to the *Jardin*, which, as is well known, lies like an island
in the middle of the Glacier du Talèfre. We reached the
place by the usual route, and found some tourists reposing
on the soft green sward which covers the lower portion, and
to which, and the flowers which spangle it, the place owes its
name. Towards the summit of the Jardin, a rock jutted
forward, apparently the very apex of the place, or at least
hiding by its prominence everything that might exist behind it; leaving our guide with the instrument, we aimed
at this, and soon left the grass and flowers behind us.
Stepping amid broken fragments of rock, along slopes of
granite, with fat felspar crystals which gave the boots a
hold, and crossing at intervals patches of snow, which
continued still to challenge the summer heat, I at length
found myself upon the peak referred to; and, although it
was not the highest, the unimpeded view which it commanded induced me to get astride it. The Jardin was completely encircled by the ice of the glacier, and this was
held in a mountain basin, which was bounded all round
by a grand and cliffy rim. The outline of the dark brown
crags—a deeply serrated and irregular line—was forcibly
drawn against the blue heaven, and still more strongly
against some white and fleecy clouds which lay here and
there behind it; while detached spears and pillars of
rock, sculptured by frost and lightning, stood like a kind of
defaced statuary along the ridge. All round the basin the

snow reared itself like a buttress against the precipitous
cliffs, being streaked and fluted by the descent of blocks
from the summits.   This mighty tub is the collector of one
of the tributaries of the Mer de Glace.   According to Professor Forbes, its greatest diameter is 4200 yards, and out
of it the half-formed ice is squeezed through a precipitous
gorge about 700 yards wide, forming there the ice cascade
of the Talèfre.   Bounded on one side by the Grande
Jorasse, and on the other by Mont Mallet, the principal
tributary of the Glacier de Léchaud lay white and pure
upon the mountain slope.   Round further to the right we
had the vast plateau whence the Glacier du Géant is fed,
fenced on the left by the Aiguille du Géant and the
Aiguille Noire, and on the right by the Monts Maudits and
Mont Blanc.   The scene was a truly majestic one.   The
mighty Aiguilles piercing the sea of air, the soft white
clouds floating here and there behind them ; the shining
snow with its striped faults and precipices; the deep blue
firmament overhead ; the peals of avalanches and the
sound of water ;—all conspired to render the scene glorious, and our enjoyment of it deep.

A voice from above hailed me as I moved from my
perch ; it was my friend, who had found a lodgment
upon the edge of a rock which was quite detached from
the Jardin, being the first to lift its head in opposition to
the descending *névé*.   Making a détour round a steep
concave slope of the glacier, I reached the flat summit of
the rock.   The end of a ridge of ice abutted against it,
which was split and bent by the pressure so as to form a
kind of arch.   I cut steps in the ice, and ascended until I got
beneath the azure roof.   Innumerable little rills of pellucid
water descended from it.   Some came straight down, clear
for a time, and apparently motionless, rapidly tapering at
first, and more slowly afterwards, until, at the point of maximum contraction, they resolved themselves into strings of

liquid pearls which pattered against the ice floor underneath. Others again, owing to the directions of the little streamlets of which they were constituted, formed spiral figures of great beauty: one liquid vein wound itself round another, forming a spiral protuberance, and owing to the centrifugal motion thus imparted, the vein, at its place of rupture, scattered itself laterally in little liquid spherules.* Even at this great elevation the structure of the ice was fairly developed, not with the sharpness to be observed lower down, but still perfectly decided. Blue bands crossed the ridge of ice to which I have referred, at right angles to the direction of the pressure.

I descended, and found my friend beneath an overhanging rock. Immediately afterwards a peal like that of thunder shook the air, and right in front of us an avalanche darted down the brown cliffs, then along a steep slope of snow which reared itself against the mountain wall, carrying with it the débris of the rocks over which it passed, until it finally lay a mass of sullied rubbish at the base of the incline: the whole surface of the Talèfre is thus soiled. Another peal was heard immediately afterwards, but the avalanche which caused it was hidden from us by a rocky promontory. From this same promontory the greater portion of the medial moraine which descends the cascade of the Talèfre is derived, forming at first a gracefully winding curve, and afterwards stretching straight to the summit of the fall. In the chasms of the cascade its boulders are engulfed, but the lost moraine is restored below the fall, as if disgorged by the ice which had swallowed it. From the extremity of the Jardin itself a mere driblet of a moraine proceeds, running parallel to the former, and like it disappearing at the summit of the cascade.

* The recent hydraulic researches of Professor Magnus furnish some beautiful illustrations of this action.

We afterwards descended towards the cascade, but long before this is attained the most experienced iceman would find himself in difficulty. Transverse crevasses are formed, which follow each other so speedily as to leave between them mere narrow ridges of ice, along which we moved cautiously, jumping the adjacent fissures, or getting round them, as the case demanded. As we approached the jaws of the gorge, the ridges dwindled to mere plates and wedges, which being bent and broken by the lateral pressure, added to the confusion, and warned us not to advance. The position was in some measure an exciting one. Our guide had never been here before; we were far from the beaten track, and the riven glacier wore an aspect of treacherous hostility. As at the base of the *séracs*, a subterranean noise sometimes announced the falling of ice-blocks into hollows underneath, the existence of which the resonant concussion of the fallen mass alone revealed. There was thus a dash of awe mingled with our thoughts; a stirring up of the feelings which troubled the coolness of the intellect. We finally swerved to the right, and by a process the reverse of straightforward reached the Couvercle. Nightfall found us at the threshold of our hotel.

(10.)

On the 5th we were engaged for some time in an important measurement at the Tacul. We afterwards ascended towards the *séracs*, and determined the inclinations of the Glacier du Géant downwards. Dense cloud-masses gathered round the points of the Aiguilles, and the thunder bellowed at intervals from the summit of Mont Blanc. As we descended the Mer de Glace the valley in front of us was filled with a cloud of pitchy darkness. Suddenly from side

to side this field of gloom was riven by a bar of lightning of intolerable splendour; it was followed by a peal of
commensurate grandeur, the echos of which leaped from
cliff to cliff long after the first sound had died away. The
discharge seemed to unlock the clouds above us, for they
showered their liquid spheres down upon us with a momentum like that of swan-shot: all the way home we were
battered by this pellet-like rain. On the 6th the rain continued with scarcely any pause; on the 7th I was engaged
all day upon the Glacier du Géant; on the morning of the
8th heavy hail had fallen there, the stones being perfect
spheres; the rounded rain-drops had solidified during their
descent without sensible change of form. When this hail
was squeezed together, it exactly resembled a mass of
oolitic limestone which I had picked up in 1853 near
Blankenburg in the Harz. Mr. Hirst and myself were engaged together this day taking the inclinations: he struck
his theodolite at the Angle, and went home accompanied by
Simond, and, the evening being extremely serene, I pursued
my way down the centre of the glacier towards the
Echelets. The crevasses as I advanced became more
deep and frequent, the ridges of ice between them becoming gradually narrower. They were very fine, their downward faces being clear cut, perfectly vertical, and in many
cases beautifully veined. Vast plates of ice moreover
often stood out midway between the walls of the chasms,
as if cloven from the glacier and afterwards set on edge.
The place was certainly one calculated to test the skill
and nerve of an iceman; and as the day drooped, and
the shadow in the valley deepened, a feeling approaching to awe took possession of me. My route was an exaggerated zigzag; right and left amid the chasms wherever
a hope of progress opened; and here I made the experience
which I have often repeated since, and laid to heart as
regards intellectual work also, that enormous difficulties

may be overcome when they are attacked in earnest. Sometimes I found myself so hedged in by fissures that escape seemed absolutely impossible; but close and resolute examination so often revealed a means of exit, that I felt in all its force the brave verity of the remark of Mirabeau, that the word "impossible" is a mere blockhead of a word. It finally became necessary to reach the shore, but I found this a work of extreme difficulty. At length, however, it became pretty evident that, if I could cross a certain crevasse, my retreat would be secured. The width of the fissure seemed to be fairly within jumping distance, and if I could have calculated on a safe purchase for my foot I should have thought little of the spring; but the ice on the edge from which I was to leap was loose and insecure, and hence a kind of nervous thrill shot through me as I made the bound. The opposite side was fairly reached, but an involuntary tremor shook me all over after I felt myself secure. I reached the edge of the glacier without further serious difficulty, and soon after found myself steeped in the creature comforts of our hotel.

On Monday, August 10th, I had the great pleasure of being joined by my friend Huxley; and though the weather was very unpromising, we started together up the glacier, he being desirous to learn something of its general features, and, if possible, to reach the Jardin. We reached the Couvercle, and squeezed ourselves through the Egralets; but here the rain whizzed past us, and dense fog settled upon the cascade of the Talèfre, obscuring all its parts. We met Mr. Galton, the African traveller, returning from an attempt upon the Jardin; and learning that his guides had lost their way in the fog, we deemed it prudent to return.

The foregoing brief notes will have informed the reader that at the period of Mr. Huxley's arrival I was not without due training upon the ice; I may also remark, that on

the 25th of July I reached the summit of the Col du
Géant, accompanied by the boy Balmat, and returned to
the Montanvert on the same day. My health was perfect,
and incessant practice had taught me the art of dealing
with the difficulties of the ice. From the time of my
arrival at the Montanvert the thought of ascending Mont
Blanc, and thus expanding my knowledge of the glaciers,
had often occurred to me, and I think I was justified in
feeling that the discipline which both my friend Hirst
and myself had undergone ought to enable us to accomplish the journey in a much more modest way than ordinary. I thought a single guide sufficient for this purpose,
and I was strengthened in this opinion by the fact that
Simond, who was a man of the strictest prudence, and who
at first declared four guides to be necessary, had lowered
his demand first to two, and was now evidently willing to
try the ascent with us alone.

On mentioning the thing to Mr. Huxley he at once
resolved to accompany us. On the 11th of August the
weather was exceedingly fine, though the snow which had
fallen during the previous days lay thick upon the glacier.
At noon we were all together at the Tacul, and the subject
of attempting Mont Blanc was mooted and discussed. My
opinion was that it would be better to wait until the fresh
snow which loaded the mountain had disappeared ; but the
weather was so exquisite that my friends thought it best to
take advantage of it. We accordingly entered into an
agreement with our guide, and immediately descended to
make preparations for commencing the expedition on the
following morning.

FIRST ASCENT OF MONT BLANC, 1857.

(11.)

ON Wednesday, the 12th of August, we rose early, after a very brief rest on my part. Simond had proposed to go down to Chamouni, and commence the ascent in the usual way, but we preferred crossing the mountains from the Montan-vert, straight to the Glacier des Bossons. At eight o'clock we started, accompanied by two porters who were to carry our provisions to the Grands Mulets. Slowly and silently we climbed the hill-side towards Charmoz. We soon passed the limits of grass and rhododendrons, and reached the slabs of gneiss which overspread the summit of the ridge, lying one upon the other like coin upon the table of a money-changer. From the highest point I turned to have a last look at the Mer de Glace ; and through a pair of very dark spectacles I could see with perfect distinct-ness the looped dirt-bands of the glacier, which to the naked eye are scarcely discernible except by twilight. Flanking our track to the left rose a series of mighty Aiguilles—the Aiguille de Charmoz, with its bent and rifted pinnacles; the Aiguille du Grepon, the Aiguille de Blaitière, the Aiguille du Midi, all piercing the heavens with their sharp pyramidal summits. Far in front of us rose the grand snow-cone of the Dôme du Gouté, while, through a forest of dark pines which gathered like a cloud at the foot of the mountain, gleamed the white minarets of the Glacier des Bossons. Below us lay the Valley of Chamouni, beyond which were the Brevent and the chain of the Aiguilles Rouges; behind us was the granite obelisk of the Aiguille du Dru, while close at hand science found a corporeal form in a pyramid of stones used as a trigonometrical station by Professor Forbes. Sound is known to travel better up hill than down, because the pulses transmitted from a denser medium to a rarer, suffer less loss of intensity than when the transmission is in the opposite direction; and now the mellow voice of the Arve came swinging upwards from the heavier air of the valley to the lighter air of the hills in rich deep cadences.

The way for a time was excessively rough, our route being overspread with the fragments of peaks which had once reared themselves to our left, but which frost and lightning had shaken to pieces, and poured in granite avalanches down the mountain. We were sometimes among huge angular boulders, and sometimes amid lighter shingle, which gave way at every step, thus forcing us to shift our footing incessantly. Escaping from these, we crossed the succession of secondary glaciers which lie at the feet of the Aiguilles, and having secured firewood found ourselves after some hours of hard work at the Pierre l'Echelle. Here we were furnished with leggings of coarse woollen cloth to keep out the snow; they were tied under the knees and quite tightly again over the insteps, so that the legs were effectually protected. We had some refreshment, possessed ourselves of the ladder, and entered upon the glacier.

The ice was excessively fissured: we crossed crevasses and crept round slippery ridges, cutting steps in the ice wherever climbing was necessary. This rendered our progress very slow. Once, with the intention of lending a helping hand, I stepped forward upon a block of granite which happened to be poised like a rocking stone upon the ice, though I did not know it; it treacherously turned under me; I fell, but my hands were in instant requisition, and I escaped with a bruise, from which, however, the blood oozed angrily. We found the ladder necessary in

crossing some of the chasms, the iron spikes at its end
being firmly driven into the ice at one side, while the
other end rested on the opposite side of the fissure. The
middle portion of the glacier was not difficult. Mounds of
ice rose beside us right and left, which were sometimes
split into high towers and gaunt-looking pyramids, while
the space between was unbroken. Twenty minutes' walking brought us again to a fissured portion of the glacier,
and here our porter left the ladder on the ice behind him.
For some time I was not aware of this, but we were soon
fronted by a chasm to pass which we were in consequence
compelled to make a long and dangerous circuit amid
crests of crumbling ice. This accomplished, we hoped
that no repetition of the process would occur, but we
speedily came to a second fissure, where it was necessary to step from a projecting end of ice to a mass of soft
snow which overhung the opposite side. Simond could
reach this snow with his long-handled axe; he beat it
down to give it rigidity, but it was exceedingly tender,
and as he worked at it he continued to express his fears
that it would not bear us. I was the lightest of the
party, and therefore tested the passage first; being partially lifted by Simond on the end of his axe, I crossed
the fissure, obtained some anchorage at the other side, and
helped the others over. We afterwards ascended until
another chasm, deeper and wider than any we had hitherto
encountered, arrested us. We walked alongside of it in
search of a snow bridge, which we at length found, but the
keystone of the arch had unfortunately given way, leaving
projecting eaves of snow at both sides, between which we
could look into the gulf, till the gloom of its deeper portions cut the vision short. Both sides of the crevasse were
sounded, but no sure footing was obtained; the snow was
beaten and carefully trodden down as near to the edge as
possible, but it finally broke away from the foot and fell

into the chasm. One of our porters was short-legged and a bad iceman; the other was a daring fellow, and he now threw the knapsack from his shoulders, came to the edge of the crevasse, looked into it, but drew back again. After a pause he repeated the act, testing the snow with his feet and staff. I looked at the man as he stood beside the chasm manifestly undecided as to whether he should take the step upon which his life would hang, and thought it advisable to put a stop to such perilous play. I accordingly interposed, the man withdrew from the crevasse, and he and Simond descended to fetch the ladder.

While they were away Huxley sat down upon the ice, with an expression of fatigue stamped upon his countenance: the spirit and the muscles were evidently at war, and the resolute will mixed itself strangely with the sense of peril and feeling of exhaustion. He had been only two days with us, and, though his strength is great, he had had no opportunity of hardening himself by previous exercise upon the ice for the task which he had undertaken. The ladder now arrived, and we crossed the crevasse. I was intentionally the last of the party, Huxley being immediately in front of me. The determination of the man disguised his real condition from everybody but myself, but I saw that the exhausting journey over the boulders and débris had been too much for his London limbs. Converting my waterproof havresack into a cushion, I made him sit down upon it at intervals, and by thus breaking the steep ascent into short stages we reached the cabin of the Grands Mulets together. Here I spread a rug on the boards, and placing my bag for a pillow, he lay down, and after an hour's profound sleep he rose refreshed and well; but still he thought it wise not to attempt the ascent farther. Our porters left us: a baton was stretched across the room over the stove, and our wet socks and leggings were thrown across it to dry; our boots were placed around the fire, and we set about preparing our evening meal. A pan was placed upon the fire, and filled with snow, which in due time melted and boiled; I ground some chocolate and placed it in the pan, and afterwards ladled the beverage into the vessels we possessed, which consisted of two earthen dishes and the metal cases of our brandy flasks. After supper Simond went out to inspect the glacier, and was observed by Huxley, as twilight fell, in a state of deep contemplation beside a crevasse.

Gradually the stars appeared, but as yet no moon. Before lying down we went out to look at the firmament, and noticed, what I suppose has been observed to some extent by everybody, that the stars near the horizon twinkled busily, while those near the zenith shone with a steady light. One large star in particular excited our admiration; it flashed intensely, and changed colour incessantly, sometimes blushing like a ruby, and again gleaming like an emerald. A determinate colour would sometimes remain constant for a sensible time, but usually the flashes followed each other in very quick succession. Three planks were now placed across the room near the stove, and upon these, with their rugs folded round them, Huxley and Hirst stretched themselves, while I nestled on the boards at the most distant end of the room. We rose at eleven o'clock, renewed the fire and warmed ourselves, after which we lay down again. I at length observed a patch of pale light upon the wooden wall of the cabin, which had entered through a hole in the end of the edifice, and rising found that it was past one o'clock. The cloudless moon was shining over the wastes of snow, and the scene outside was at once wild, grand, and beautiful.

Breakfast was soon prepared, though not without difficulty; we had no candles, they had been forgotten; but I fortunately possessed a box of wax matches, of which Huxley took charge, patiently igniting them in succession, and thus giving us a tolerably continuous light. We had some tea, which had been made at the Montanvert, and carried to the Grands Mulets in a bottle. My memory of that tea is not pleasant; it had been left a whole night in contact with its leaves, and smacked strongly of tannin. The snow-water, moreover, with which we diluted it was not pure, but left a black residuum at the bottom of the dishes in which the beverage was served. The few provisions deemed necessary being placed in Simond's knapsack, at twenty minutes past two o'clock we scrambled down the rocks, leaving Huxley behind us.

The snow was hardened by the night's frost, and we were cheered by the hope of being able to accomplish the ascent with comparatively little labour. We were environed by an atmosphere of perfect purity; the larger stars hung like gems above us, and the moon, about half full, shone with wondrous radiance in the dark firmament. One star in particular, which lay eastward from the moon, suddenly made its appearance above one of the Aiguilles, and burned there with unspeakable splendour. We turned once towards the Mulets, and saw Huxley's form projected against the sky as he stood upon a pinnacle of rock; he gave us a last wave of the hand and descended, while we receded from him into the solitudes.

The evening previous our guide had examined the glacier for some distance, his progress having been arrested by a crevasse. Beside this we soon halted: it was spanned at one place by a bridge of snow, which was of too light a structure to permit of Simond's testing it alone; we therefore paused while our guide uncoiled a rope and tied us all together. The moment was to me a peculiarly solemn one. Our little party seemed so lonely and so small amid the silence and the vastness of the surrounding scene. We were about to try our strength under unknown conditions, and as the various possibilities of the enterprise crowded on the imagination, a sense of responsibility for a moment oppressed me. But as I looked aloft and saw the glory of the heavens, my heart lightened, and I remarked cheerily to Hirst that Nature seemed to smile upon our work. "Yes," he replied, in a calm and earnest voice, "and, God willing, we shall accomplish it."

A pale light now overspread the eastern sky, which increased, as we ascended, to a daffodil tinge; this afterwards heightened to orange, deepening at one extremity into red, and fading at the other into a pure ethereal hue to which it would be difficult to assign a special name. Higher up the sky was violet, and this changed by insensible degrees into the darkling blue of the zenith, which had to thank the light of moon and stars alone for its existence. We wound steadily for a time through valleys of ice, climbed white and slippery slopes, crossed a number of crevasses, and after some time found ourselves beside a chasm of great depth and width, which extended right and left as far as we could see. We turned to the left, and marched along its edge in search of a *pont;* but matters became gradually worse : other crevasses joined on to the first one, and the further we proceeded the more riven and dislocated the ice became. At length we reached a place where further advance was impossible. Simond in his difficulty complained of the want of light, and wished us to wait for the advancing day; I, on the contrary, thought that we had light enough and ought to make use of it. Here the thought occurred to me that Simond, having been only once before to the top of the mountain, might not be quite clear about the route; the glacier, however, changes within certain limits from year to year, so that a general knowledge was all that could be expected, and we trusted to our own muscles to make good any mistake in the way of guidance. We now turned and retraced our steps along the edges of chasms where the ice was disintegrated and insecure, and succeeded at length in finding a bridge which bore us across the crevasse. This error caused us the loss of an hour, and after walking for this time we could cast a stone from the point we had attained to the place whence we had been compelled to return.

Our way now lay along the face of a steep incline of snow, which was cut by the fissure we had just passed, in a direction parallel to our route. On the heights to our right, loose ice-crags seemed to totter, and we passed two tracks over which the frozen blocks had rushed some short time previously. We were glad to get out of the range of these terrible projectiles, and still more so to escape the vicinity of that ugly crevasse. To be killed in the open air would be a luxury, compared with having the life squeezed out of one in the horrible gloom of these chasms. The blush of the coming day became more and more intense; still the sun himself did not appear, being hidden from us by the peaks of the Aiguille du Midi, which were drawn clear and sharp against the brightening sky. Right under this Aiguille were heaps of snow smoothly rounded and constituting a portion of the sources whence the Glacier du Géant is fed; these, as the day advanced, bloomed with a rosy light. We reached the Petit Plateau, which we found covered with the remains of ice avalanches; above us upon the crest of the mountain rose three mighty bastions, divided from each other by deep vertical rents, with clean smooth walls, across which the lines of annual bedding were drawn like courses of masonry. From these, which incessantly renew themselves, and from the loose and broken ice-crags near them, the boulders amid which we now threaded our way had been discharged. When they fall their descent must be sublime.

The snow had been gradually getting deeper, and the ascent more wearisome, but superadded to this at the Petit Plateau was the uncertainty of the footing between the

blocks of ice. In many places the space was merely
covered by a thin crust, which, when trod upon, instantly
yielded, and we sank with a shock sometimes to the hips.
Our way next lay up a steep incline to the Grand Plateau,
the depth and tenderness of the snow augmenting as we
ascended. We had not yet seen the sun, but, as we
attained the brow which forms the entrance to the Grand
Plateau, he hung his disk upon a spike of rock to our
left, and, surrounded by a glory of interference spectra
of the most gorgeous colours, blazed down upon us. On
the Grand Plateau we halted and had our frugal refreshment. At some distance to our left was the crevasse into
which Dr. Hamel's three guides were precipitated by an
avalanche in 1820; they are still entombed in the ice,
and some future explorer may perhaps see them disgorged
lower down, fresh and undecayed. They can hardly reach
the surface until they pass the snow-line of the glacier, for
above this line the quantity of snow that annually falls
being in excess of the quantity melted, the tendency would
be to make the ice-covering above them thicker. But it
is also possible that the waste of the ice underneath may
have brought the bodies to the bed of the glacier, where
their very bones may have been ground to mud by an
agency which the hardest rocks cannot withstand.

As the sun poured his light upon the Plateau the little
snow-facets sparkled brilliantly, sometimes with a pure
white light, and at others with prismatic colours. Contrasted with the white spaces above and around us were
the dark mountains on the opposite side of the valley of
Chamouni, around which fantastic masses of cloud were
beginning to build themselves. Mont Buet, with its cone
of snow, looked small, and the Brevent altogether mean;
the limestone bastions of the Fys, however, still presented a front of gloom and grandeur. We traversed
the Grand Plateau, and at length reached the base of an

extremely steep incline which stretched upwards towards the Corridor. Here, as if produced by a fault, consequent upon the sinking of the ice in front, rose a vertical precipice, from the coping of which vast stalactites of ice depended. Previous to reaching this place I had noticed a haggard expression upon the countenance of our guide, which was now intensified by the prospect of the ascent before him. Hitherto he had always been in front, which was certainly the most fatiguing position. I felt that I must now take the lead, so I spoke cheerily to the man and placed him behind me. Marking a number of points upon the slope as resting places, I went swiftly from one to the other. The surface of the snow had been partially melted by the sun and then refrozen, thus forming a superficial crust, which bore the weight up to a certain point, and then suddenly gave way, permitting the leg to sink to above the knee. The shock consequent on this, and the subsequent effort necessary to extricate the leg, were extremely fatiguing. My motion was complained of as too quick, and my tracks as imperfect; I moderated the former, and, to render my footholes broad and sure, I stamped upon the frozen crust, and twisted my legs in the soft mass underneath,—a terribly exhausting process. I thus led the way to the base of the Rochers Rouges, up to which the fault already referred to had prolonged itself as a crevasse, which was roofed at one place by a most dangerous-looking snow-bridge. Simond came to the front ; I drew his attention to the state of the snow, and proposed climbing the Rochers Rouges ; but, with a promptness unusual with him, he replied that this was impossible ; the bridge was our only means of passing, and we must try it. We grasped our ropes, and dug our feet firmly into the snow to check the man's descent if the *pont* gave way, but to our astonishment it bore him, and bore us safely after him. The slope which we had now to ascend had the snow swept from its surface, and was therefore firm ice. It was most dangerously steep, and, its termination being the fretted coping of the precipice to which I have referred, if we slid downwards we should shoot over this and be dashed to pieces upon the ice below.* Simond, who had come to the front to cross the crevasse, was now engaged in cutting steps, which he made deep and large, so that they might serve us on our return. But the listless strokes of his axe proclaimed his exhaustion ; so I took the implement out of his hands, and changed places with him. Step after step was hewn, but the top of the Corridor appeared ever to recede from us. Hirst was behind unoccupied, and could thus turn his thoughts to the peril of our position : he *felt* the angle on which we hung, and saw the edge of the precipice, to which less than a quarter of a minute's slide would carry us, and for the first time during the journey he grew giddy. A cigar which he lighted for the purpose tranquilized him.

I hewed sixty steps upon this slope, and each step had cost a minute, by Hirst's watch. The Mur de la Côte was still before us, and on this the guide-books informed us two or three hundred steps were sometimes found necessary. If sixty steps cost an hour, what would be the cost of two hundred ? The question was disheartening in the extreme, for the time at which we had calculated on reaching the summit was already passed, while the chief difficulties remained unconquered. Having hewn our way along the harder ice we reached snow. I again resorted to stamping to secure a footing, and while thus engaged became, for the first time, aware of the drain of force to which I was subjecting myself. The thought of being

---

* Those acquainted with the mountain will at once recognise the grave error here committed. In fact, on starting from the Grands Mulets we had crossed the glacier too far, and throughout were much too close to the Dôme du Goûté.

absolutely exhausted had never occurred to me, and from first to last I had taken no care to husband my strength. I always calculated that the *will* would serve me even should the muscles fail, but I now found that mechanical laws rule man in the long run; that no effort of will, no power of spirit, can draw beyond a certain limit upon muscular force. The soul, it is true, can stir the body to action, but its function is to excite and apply force, and not to create it.

While stamping forward through the frozen crust I was compelled to pause at short intervals; then would set out again apparently fresh, to find, however, in a few minutes that my strength ·was gone, and that I required to rest once more. In this way I gained the summit of the Corridor, when Hirst came to the front, and I felt some relief in stepping slowly after him, making use of the holes into which his feet had sunk. He thus led the way to the base of the Mur de la Côte, the thought of which had so long cast a gloom upon us; here we left our rope behind us, and while pausing I asked Simond whether he did not feel a desire to go to the summit—" *Bien sur,*" was his reply, "*mais!*" Our guide's mind was so constituted that the "*mais*" seemed essential to its peace. I stretched my hand towards him, and said, "Simond, we must do it." One thing alone I felt could defeat us: the usual time of the ascent had been more than doubled, the day was already far spent, and if the ascent would throw our subsequent descent into night it could not be contemplated.

We now faced the Mur, which was by no means so bad as we had expected. Driving the iron claws of our boots into the scars made by the axe, and the spikes of our batons into the slope above our feet, we ascended steadily until the summit was attained, and the top of the mountain rose clearly above us. We congratulated ourselves upon this; but Simond, probably fearing that our joy might become too full, remarked, " *Mais le sommet est en-*

core bien loin!" It was, alas! too true. The snow became soft again, and our weary limbs sank in it as
before. Our guide went on in front, audibly muttering
his doubts as to our ability to reach the top, and at length
he threw himself upon the snow, and exclaimed, "*Il faut
le renoncer!*" Hirst now undertook the task of rekindling
the guide's enthusiasm, after which Simond rose, exclaiming, "*Ah! comme ça me fait mal aux genoux,*" and
went forward. Two rocks break through the snow between
the summit of the Mur and the top of the mountain;
the first is called the Petits Mulets, and the highest the
Derniers Rochers. At the former of these we paused to
rest, and finished our scanty store of wine and provisions.
We had not a bit of bread nor a drop of wine left; our
brandy flasks were also nearly exhausted, and thus we had
to contemplate the journey to the summit, and the subsequent descent to the Grands Mulets, without the slightest
prospect of physical refreshment. The almost total loss
of two nights' sleep, with two days' toil superadded, made
me long for a few minutes' doze, so I stretched myself
upon a composite couch of snow and granite, and immediately fell asleep. My friend, however, soon aroused me.
"You quite frighten me," he said; "I have listened for
some minutes, and have not heard you breathe once." I
had, in reality, been taking deep draughts of the mountain air, but so silently as not to be heard.

I now filled our empty wine-bottle with snow and placed
it in the sunshine, that we might have a little water on
our return. We then rose; it was half-past two o'clock;
we had been upwards of twelve hours climbing, and I calculated that, whether we reached the summit or not, we
could at all events work *towards* it for another hour. To
the sense of fatigue previously experienced, a new phenomenon was now added—the beating of the heart. We
were incessantly pulled up by this, which sometimes be-

came so intense as to suggest danger. I counted the number of paces which we were able to accomplish without resting, and found that at the end of every twenty, sometimes at the end of fifteen, we were compelled to pause. At each pause my heart throbbed audibly, as I leaned upon my staff, and the subsidence of this action was always the signal for further advance. My breathing was quick, but light and unimpeded. I endeavoured to ascertain whether the hip-joint, on account of the diminished atmospheric pressure, became loosened, so as to throw the weight of the leg upon the surrounding ligaments, but could not be certain about it. I also sought a little aid and encouragement from philosophy, endeavouring to remember what great things had been done by the accumulation of small quantities, and I urged upon myself that the present was a case in point, and that the summation of distances twenty paces each must finally place us at the top. Still the question of time left the matter long in doubt, and until we had passed the Derniers Rochers we worked on with the stern indifference of men who were doing their duty, and did not look to consequences. Here, however, a gleam of hope began to brighten our souls; the summit became visibly nearer, Simond showed more alacrity; at length success became certain, and at half-past three P.M. my friend and I clasped hands upon the top.

The summit of the mountain is an elongated ridge, which has been compared to the back of an ass. It was perfectly manifest that we were dominant over all other mountains; as far as the eye could range Mont Blanc had no competitor. The summits which had looked down upon us in the morning were now far beneath us. The Dôme du Gouté, which had held its threatening *séracs* above us so long, was now at our feet. The Aiguille du Midi, Mont Blanc du Tacul, and the Monts Maudits, the

Talèfre with its surrounding peaks, the Grand Jorasse, Mont
Mallet, and the Aiguille du Géant, with our own familiar
glaciers, were all below us. And as our eye ranged over
the broad shoulders of the mountain, over ice hills and
valleys, plateaux and far-stretching slopes of snow, the
conception of its magnitude grew upon us, and impressed
us more and more.

The clouds were very grand—grander indeed than anything I had ever before seen. Some of them seemed to
hold thunder in their breasts, they were so dense and dark;
others, with their faces turned sunward, shone with the
dazzling whiteness of the mountain snow; while others
again built themselves into forms resembling gigantic elm
trees, loaded with foliage. Towards the horizon the luxury
of colour added itself to the magnificent alternation of
light and shade. Clear spaces of amber and ethereal green
embraced the red and purple cumuli, and seemed to form
the cradle in which they swung. Closer at hand squally
mists, suddenly engendered, were driven hither and thither
by local winds; while the clouds at a distance lay " like
angels sleeping on the wing," with scarcely visible motion.
Mingling with the clouds, and sometimes rising above them,
were the highest mountain heads, and as our eyes wandered
from peak to peak, onwards to the remote horizon, space
itself seemed more vast from the manner in which the
objects which it held were distributed.

I wished to repeat the remarkable experiment of
De Saussure upon sound, and for this purpose had requested
Simond to bring a pistol from Chamouni; but in the multitude of his cares he forgot it, and in lieu of it my host at
the Montanvert had placed in two tin tubes, of the same
size and shape, the same amount of gunpowder, securely
closing the tubes afterwards, and furnishing each of them
with a small lateral aperture. We now planted one of
them upon the snow, and bringing a strip of amadou into

communication with the touchhole, ignited its most distant end: it failed; we tried again, and were successful, the explosion tearing asunder the little case which contained the powder. The sound was certainly not so great as I should have expected from an equal quantity of powder at the sea level.*

The snow upon the summit was indurated, but of an exceedingly fine grain, and the beautiful effect already referred to as noticed upon the Stelvio was strikingly manifest. The hole made by driving the baton into the snow was filled with a delicate blue light; and, by management, its complementary pinky yellow could also be produced. Even the iron spike at the end of the baton made a hole sufficiently deep to exhibit the blue colour, which certainly depends on the size and arrangement of the snow crystals. The firmament above us was without a cloud, and of a darkness almost equal to that which surrounded the moon at 2 A.M. Still, though the sun was shining, a breeze, whose tooth had been sharpened by its passage over the snow-fields, searched us through and through. The day was also waning, and, urged by the warnings of our ever prudent guide, we at length began the descent.

Gravity was now in our favour, but gravity could not entirely spare our wearied limbs, and where we sank in the snow we found our downward progress very trying. I suffered from thirst, but after we had divided the liquefied snow at the Petits Mulets amongst us we had nothing to drink. I crammed the clean snow into my mouth, but the process of melting was slow and tantalizing to a parched

---

* I fired the second case in a field in Hampshire, and, as far as my memory enabled me to make the comparison, found its sound considerably *denser*, if I may use the expression. In 1859 I had a pistol fired at the summit of Mont Blanc: its sound was sensibly feebler and *shorter* than in the valley; it resembled somewhat the discharge of a cork from a champagne bottle, though much louder, but it could not be at all compared to the sound of a common cracker.

throat, while the chill was painful to the teeth. We marched
along the Corridor, and crossed cautiously the perilous slope
on which we had cut steps in the morning, breathing more
freely after we had cleared the ice-precipice before described. Along the base of this precipice we now wound,
diverging from our morning's track, in order to get surer
footing in the snow; it was like flour, and while descending
to the Grand Plateau we sometimes sank in it nearly to
the waist. When I endeavoured to squeeze it, so as to fill
my flask, it at first refused to cling together, behaving
like so much salt; the heat of the hand, however, soon
rendered it a little moist, and capable of being pressed into
compact masses. The sun met us here with extraordinary
power; the heat relaxed my muscles, but when fairly
immersed in the shadow of the Dôme du Gouté, the coolness
restored my strength, which augmented as the evening
advanced. Simond insisted on the necessity of haste, to
save us from the perils of darkness. "*On peut perir*" was
his repeated admonition, and he was quite right. We
reached the region of *ponts*, more weary, but, in compensation, more callous, than we had been in the morning, and
moved over the soft snow of the bridges as if we had been
walking upon eggs. The valley of Chamouni was filled with
brown-red clouds, which crept towards us up the mountain;
the air around and above us was, however, clear, and the
chastened light told us that day was departing. Once as
we hung upon a steep slope, where the snow was exceedingly soft, Hirst omitted to make his footing sure; the soft
mass gave way, and he fell, uttering a startled shout as he
went down the declivity. I was attached to him, and, fixing
my feet suddenly in the snow, endeavoured to check his
fall, but I seemed a mere feather in opposition to the force
with which he descended.* I fell, and went down after him;
and we carried quite an avalanche of snow along with us,

* I believe that I could stop him now (1860).

in which we were almost completely hidden at the bottom
of the slope. All further dangers, however, were soon
past, and we went at a headlong speed to the base of the
Grand Mulets; the sound of our batons against the rocks
calling Huxley forth. A position more desolate than his
had been can hardly be imagined. For seventeen hours
he had been there. He had expected us at two o'clock in
the afternoon; the hours came and passed, and till seven
in the evening he had looked for us. "To the end of my
life," he said, "I shall never forget the sound of those
batons." It was his turn now to nurse me, which he did,
repaying my previous care of him with high interest. We
were all soon stretched, and, in spite of cold and hard
boards, I slept at intervals; but the night, on the whole,
was a weary one, and we rose next morning with muscles
more tired than when we lay down.

Friday, 14*th August.*—Hirst was almost blind this morning; and our guide's eyes were also greatly inflamed.
We gathered our things together, and bade the Grands
Mulets farewell. It had frozen hard during the night,
and this, on the steeper slopes, rendered the footing very
insecure. Simond, moreover, appeared to be a little bewil-
dered, and I sometimes preceded him in cutting the
steps, while Hirst moved among the crevasses like a blind
man; one of us keeping near him, so that he might feel for
the actual places where our feet had rested, and place his own
in the same position. It cost us three hours to cross from
the Grands Mulets to the Pierre l'Echelle, where we discarded our leggings, had a mouthful of food, and a brief
rest. Once upon the safe earth Simond's powers seemed to
be restored, and he led us swiftly downwards to the little
auberge beside the Cascade du Tard, where we had some
excellent lemonade, equally choice cognac, fresh strawberries and cream. How sweet they were, and how beautiful we thought the peasant girl who served them! Our

guide kept a little hotel, at which we halted, and found it clean and comfortable. We were, in fact, totally unfit to go elsewhere. My coat was torn, holes were kicked through my boots, and I was altogether ragged and shabby. A warm bath before dinner refreshed all mightily. Dense clouds now lowered upon Mont Blanc, and we had not been an hour at Chamouni when the breaking up of the weather was announced by a thunder-peal. We had accomplished our journey just in time.

(12.)

After our return we spent every available hour upon the ice, working at questions which shall be treated under their proper heads, each day's work being wound up by an evening of perfect enjoyment. Roast mutton and fried potatos were our incessant fare, for which, after a little longing for a change at first, we contracted a final and permanent love. As the year advanced, moreover, and the grass sprouted with augmented vigour on the slopes of the Montanvert, the mutton, as predicted by our host, became more tender and juicy. We had also some capital Sallenches beer, cold as the glacier water, but effervescent as champagne. Such were our food and drink. After dinner we gathered round the pine-fire, and I can hardly think it possible for three men to be more happy than we then were. It was not the goodness of the conversation, nor any high intellectual element, which gave the charm to our gatherings; the gladness grew naturally out of our own perfect health, and out of the circumstances of our position. Every fibre seemed a repository of latent joy, which the slightest stimulus sufficed to bring into conscious action.

On the 17th I penetrated with Simond through thick gloom to the Tacul; on the 18th we set stakes at the same

place: on the same day, while crossing the medial moraine
of the Talèfre, a little below the cascade, a singular noise
attracted my attention; it seemed at first as if a snake
were hissing about my feet. On changing my position the
sound suddenly ceased, but it soon recommenced. There
was some snow upon the glacier, which I removed, and
placed my ear close to the ice, but it was difficult
to fix on the precise spot from which the sound issued.
I cut away the disintegrated portion of the surface, and
at length discovered a minute crack, from which a
stream of air issued, which I could feel as a cold blast
against my hand. While cutting away the surface further, I stopped the little "blower." A marmot screamed
near me, and while I paused to look at the creature
scampering up the crags, the sound commenced again,
changing its note variously—hissing like a snake, singing like a kettle, and sometimes chirruping intermittently
like a bird. On passing my fingers to and fro over
the crack, I obtained a succession of audible puffs; the
current was sufficiently strong to blow away the corner of
my gauze veil when held over the fissure. Still the crack
was not wide enough to permit of the entrance of my
finger nail; and to issue with such force from so minute a
rent the air must have been under considerable pressure.
The origin of the blower was in all probability the following:—When the ice is recompacted after having descended
a cascade, it is next to certain that chambers of air will be
here and there enclosed, which, being powerfully squeezed
afterwards, will issue in the manner described whenever a
crack in the ice furnishes it with a means of escape. In
my experiments on flowing mud, for example, the air
entrapped in the mass while descending from the sluice
into the trough, bursts in bubbles from the surface at
a short distance downwards.

I afterwards examined the Talèfre cascade from summit

to base, with reference to the structure, until at the close of the day thickening clouds warned me off. I went down the glacier at a trot, guided by the boulders capped with little cairns which marked the route. The track which I had pursued for the last five weeks amid the crevasses near l'Angle was this day barely passable. The glacier had changed, my work was drawing to a close, and, as I looked at the objects which had now become so familiar to me, I felt that, though not viscous, the ice did not lack the quality of "adhesiveness," and I felt a little sad at the thought of bidding it so soon farewell.

At some distance below the Montanvert the Mer de Glace is riven from side to side by transverse crevasses: these fissures indicate that the glacier where they occur is in a state of longitudinal strain which produces transverse fracture. I wished to ascertain the amount of stretching which the glacier here demanded, and which the ice was not able to give; and for this purpose desired to compare the velocity of a line set out across the fissured portion with that of a second line staked out across the ice before it had become thus fissured. A previous inspection of the glacier through the telescope of our theodolite induced us to fix on a place which, though much riven, still did not exclude the hope of our being able to reach the other side. Each of us was, as usual, armed with his own axe; and carrying with us suitable stakes, my guide and myself entered upon this portion of the glacier on the morning of the 19th of August.

I was surprised on entering to find some veins of white ice, which from their position and aspect appeared to be derived from the Glacier du Géant; but to these I shall subsequently refer. Our work was extremely difficult; we penetrated to some distance along one line, but were finally forced back, and compelled to try another. Right and left of us were profound fissures, and once

a cone of ice forty feet high leaned quite over our track.
In front of us was a second leaning mass borne by a
mere stalk, and so topheavy that one wondered why the
slight pedestal on which it rested did not suddenly crack
across. We worked slowly forwards, and soon found ourselves in the shadow of the topheavy mass above referred
to; and from which I escaped with a wounded hand,
caused by over-haste. Simond surmounted the next
ridge and exclaimed, "*Nous nous trouverons perdus!*" I
reached his side, and on looking round the place saw that
there was no footing for man. The glacier here, as shown
in the frontispiece, was cut up into thin wedges, separated
from each other by profound chasms, and the wedges were
so broken across as to render creeping along their edges
quite impossible. Thus brought to a stand, I fixed a
stake at the point where we were forced to halt, and
retreated along edges of detestable granular ice, which fell
in showers into the crevasses when struck by the axe. At
one place an exceedingly deep fissure was at our left,
which was joined, at a sharp angle, by another at our
right, and we were compelled to cross at the place of
intersection : to do this we had to trust ourselves to a
projecting knob of that vile rotten ice which I had learned
to fear since my experience of it on the Col du Géant.
We finally escaped, and set out our line at another place,
where the glacier, though badly cut, was not impassable.

On the 20th we made a series of final measurements at
the Tacul, and determined the motion of two lines which
we had set out upon the previous day. On the 21st we
quitted the Montanvert; I had been there from the 15th
of July, and the longer I remained the better I liked the
establishment and the people connected with it. It was
then managed by Joseph Tairraz and Jules Charlet, both
of whom showed us every attention. In 1858 and 1859 I
had occasion to revisit the establishment, which was then

managed by Jules and his brother, and found in it the
same good qualities. During my winter expedition of
1859 I also found the same readiness to assist me in every
possible way ; honest Jules expressing his willingness to
ascend through the snow to the auberge if I thought his
presence would in any degree contribute to my comfort.

We crossed the glacier, and descended by the Chapeau
to the Cascade des Bois, the inclination of which and of
the lower portion of the glacier we then determined. The
day was magnificent. Looking upwards, the Aiguilles de
Charmoz and du Dru rose right and left like sentinels of
the valley, while in front of us the ice descended the steep,
a bewildering mass of crags and chasms. At the other
side was the pine-clad slope of the Montanvert. Further
on the Aiguille du Midi threw its granite pyramid between
us and Mont Blanc; on the Dôme du Gouté the *sérac* of
the mountain were to be seen, while issuing as if from a
cleft in the mountain side the Glacier des Bossons thrust
through the black pines its snowy tongue. Below us was
the beautiful valley of Chamouni itself, through which
the Arve and Arveiron rushed like enlivening spirits. We
finally examined a grand old moraine produced by a Mer
de Glace of other ages, when the ice quite crossed the
valley of Chamouni and abutted against the opposite mountain-wall.

Simond had proved himself a very valuable assistant;
he was intelligent and perfectly trustworthy ; and though
the peculiar nature of my work sometimes caused me
to attempt things against which his prudence protested, he
lacked neither strength nor courage. On reaching Chamouni and adding up our accounts, I found that I had not
sufficient cash to pay him ; money was waiting for me at
the post-office in Geneva, and thither it was arranged that
my friend Hirst should proceed next morning, while I was
to await the arrival of the money at Chamouni. My guide

heard of this arrangement, and divined its cause : he came to me, and in the most affectionate manner begged of me to accept from him the loan of 500 francs. Though I did not need the loan, the mode in which it was offered to me augmented the kindly feelings which I had long entertained towards Simond, and I may add that my intercourse with him since has served merely to confirm my first estimate of his worthiness.

EXPEDITION OF 1858.

(13.)

I HAD confined myself during the summer of 1857 to the Mer de Glace and its tributaries, desirous to make my knowledge accurate rather than extensive. I had made the acquaintance of all accessible parts of the glacier, and spared no pains to master both the details and the meaning of the laminated structure of the ice, but I found no fact upon which I could take my stand and say to an advocate of an opposing theory, "This is unassailable." In experimental science we have usually the power of changing the conditions at pleasure; if Nature does not reply to a question we throw it into another form; a combining of conditions is, in fact, the essence of experiment. To meet the requirements of the present question, I could not twist the same glacier into various shapes, and throw it into different states of strain and pressure; but I might, by visiting many glaciers, find all needful conditions fulfilled in detail, and by observing these I hoped to confer upon the subject the character and precision of a true experimental inquiry.

The summer of 1858 was accordingly devoted to this purpose, when I had the good fortune to be accompanied by Professor Ramsay, the author of some extremely interesting papers upon ancient glaciers. Taking Zürich, Schaffhausen, and Lucerne in our way, we crossed the Brünig on the 22nd of July, and met my guide, Christian Lauener, at Meyringen. On the 23rd we visited the glacier of Rosenlaui, and the glacier of the Schwartzwald, and reached Grindelwald in the evening of the same day. My expedition with Mr. Huxley had taught me that the Lower Grindelwald Glacier was extremely instructive, and I was anxious to see many parts of it once more; this I did, in company with Ramsay, and we also spent a day upon the upper glacier, after which our path lay over the Strahleck to the glaciers of the Aar and of the Rhone.

PASSAGE OF THE STRAHLECK.

(14.)

ON Monday, the 26th of July, we were called at 4 A.M., and found the weather very unpromising, but the two mornings which preceded it had also been threatening without any evil result. There was, it is true, something more than usually hostile in the aspect of the clouds which sailed sullenly from the west, and smeared the air and mountains as if with the dirty smoke of a manufacturing town. We despatched our coffee, went down to the bottom of the Grindelwald valley, up the opposite slope, and were soon amid the gloom of the pines which partially cover it. On emerging from these, a watery gleam on the mottled head of the Eiger was the only evidence of direct sunlight in that direction. To our left was the Wetterhorn surrounded by wild and disorderly clouds, through the fissures of which the morning light glared strangely. For a time the Heisse Platte was seen, a dark brown patch amid the ghastly blue which overspread the surrounding slopes of snow. The clouds once rolled up, and revealed for a moment the summits of the Viescherhörner; but they immediately settled down again, and hid the mountains from top to base. Soon afterwards they drew themselves partially aside, and a patch of blue over the Strahleck gave us hope and pleasure. As we ascended, the

prospect in front of us grew better, but that behind us—
and the wind came from behind—grew worse. Slowly and
stealthily the dense neutral-tint masses crept along the
sides of the mountains, and seemed to dog us like spies;
while over the glacier hung a thin veil of fog, through which
gleamed the white minarets of the ice.

When we first spoke of crossing the Strahleck, Lauener
said it would be necessary to take two guides at least; but
after a day's performance on the ice he thought we might
manage very well by taking, in addition to himself, the
herd of the alp, over the more difficult part of the pass.
He had further experience of us on the second day,
and now, as we approached the herd's hut, I was amused
to hear him say that he thought any assistance beside his
own unnecessary. Relying upon ourselves, therefore, we
continued our route, and were soon upon the glacier, which
had been rendered smooth and slippery through the removal of its disintegrated surface by the warm air. Crossing the Srahleck branch of the glacier to its left side, we
climbed the rocks to the grass and flowers which clothe the
slopes above them. Our way sometimes lay over these, sometimes along the beds of streams, across turbulent brooks,
and once around the face of a cliff, which afforded us
about an inch of ledge to stand upon, and some protruding splinters to lay hold of by the hands. Having reached
a promontory which commanded a fine view of the glacier,
and of the ice cascade by which it was fed, I halted, to
check the observations already made from the side of the
opposite mountain. Here, as there, cliffy ridges were seen
crossing the cascade of the glacier, with interposed spaces
of dirt and débris—the former being toned down, and the
latter squeezed towards the base of the fall, until finally
the ridges swept across the glacier, in gentle swellings, from
side to side ; while the valleys between them, holding the
principal share of the superficial impurity, formed the

cradles of the so-called Dirt-Bands. These swept concentric with the protuberances across the glacier, and
remained upon its surface even after the swellings had
disappeared. The swifter flow of the centre of the glacier
tends of course incessantly to lengthen the loops of the
bands, and to thrust the summits of the curves which they
form more and more in advance of their lateral portions.
The depressions between the protuberances appeared to
be furrowed by minor wrinkles, as if the ice of the depressions had yielded more than that of the protuberances.
This, I think, is extremely probable, though it has never
yet been proved. Three stakes, placed, one on the summit,
another on the frontal slope, and another at the base of a
protuberance, would, I think, move with unequal velocities.
They would, I think, shew that, upon the large and general motion of the glacier, smaller motions are superposed,
as minor oscillations are known to cover parasitically the
large ones of a vibrating string. Possibly, also, the dirt-
bands may owe something to the squeezing of impuri-
ties out of the glacier to its surface in the intervals
between the swellings. From our present position we
could also see the swellings on the Viescherhörner branch
of the glacier, in the valleys between which coarse shingle
and débris were collected, which would form dirt-bands if
they could. On neither branch, however, do the bands
attain the definition and beauty which they possess upon
the Mer de Glace.

After an instructive lesson we faced our task once more,
passing amid crags and boulders, and over steep moraines,
from which the stones rolled down upon the slightest disturbance. While crossing a slope of snow with an inclination of 45°, my footing gave way, I fell, but turned promptly
on my face, dug my staff deeply into the snow, and arrested
the motion before I had slid a dozen yards. Ramsay
was behind me, speculating whether he should be able to

pass the same point without slipping; before he reached
it, however, the snow yielded, he fell, and slid swiftly
downwards. Lauener, whose attention had been aroused
by my fall, chanced to be looking round when Ramsay's
footing yielded. With the velocity of a projectile he threw
himself upon my companion, seized him, and brought him
to rest before he had reached the bottom of the slope. The
act made a very favourable impression upon me, it was so
prompt and instinctive. An eagle could not swoop upon its
prey with more directness of aim and swiftness of execution.

While this went on the clouds were playing hide and
seek with the mountains. The ice-crags and pinnacles to our
left, looming through the haze, seemed of gigantic proportions, reminding one of the Hades of Byron's ' Cain.'

"How sunless and how vast are these dim realms!"

We climbed for some time along the moraine which
flanks the cascade, and on reaching the level of the brow
Lauener paused, cast off his knapsack, and declared for
breakfast. While engaged with it the dense clouds which
had crammed the gorge and obscured the mountains, all
melted away, and a scene of indescribable magnificence
was revealed. Overhead the sky suddenly deepened to
dark blue, and against it the Finsteraarhorn projected his
dark and mighty mass. Brown spurs jutted from the
mountain, and between them were precipitous snow-slopes,
fluted by the descent of rocks and avalanches, and broken
into ice-precipices lower down. Right in front of us, and
from its proximity more gigantic to the eye, was the
Shreckhorn, while from couloirs and mountain-slopes
the matter of glaciers yet to be was poured into the
vast basin on the rim of which we now stood.

This it was next our object to cross; our way lying
in part through deep snow-slush, the scene changing
perpetually from blue heaven to gray haze which massed
itself at intervals in dense clouds about the moun-

tains. After crossing the basin our way lay partly over
slopes of snow, partly over loose shingle, and at one place
along the edge of a formidable precipice of rock. We sat
down sometimes to rest, and during these pauses, though
they were very brief, the scene had time to go through
several of its Protean mutations. At one moment all would
be perfectly serene, no cloud in the transparent air to tell
us that any portion of it was in motion, while the blue
heaven threw its flattened arch over the magnificent amphitheatre. Then in an instant, from some local cauldron,
the vapour would boil up suddenly, eddying wildly in the
air, which a moment before seemed so still, and enveloping the entire scene. Thus the space enclosed by the
Finsteraarhorn, the Viescherhörner, and the Shreckhorn,
would at one moment be filled with fog to the mountain
heads, every trace of which a few minutes sufficed to sweep
away, leaving the unstained blue of heaven behind it,
and the mountains showing sharp and jagged outlines
in the glassy air. One might be almost led to imagine
that the vapour molecules endured a strain similar to
that of water cooled below its freezing point, or heated
beyond its boiling point; and that, on the strain being relieved by the sudden yielding of the opposing force, the
particles rushed together, and thus filled in an instant the
clear atmosphere with aqueous precipitation.

I had no idea that the Strahleck was so fine a pass.
Whether it is the quality of my mind to take in the glory
of the present so intensely as to make me forgetful of the
glory of the past, I know not, but it appeared to me that I
had never seen anything finer than the scene from the
summit. The amphitheatre formed by the mountains
seemed to me of exceeding magnificence; nor do I think
that my feeling was subjective merely; for the simple
magnitude of the masses which built up the spectacle
would be sufficient to declare its grandeur. Looking down

F

towards the Glacier of the Aar, a scene of wild beauty
and desolation presented itself. Not a trace of vegetation
could be seen along the whole range of the bounding
mountains; glaciers streamed from their shoulders into
the valley beneath, where they welded themselves to form
the Finsteraar affluent of the Unteraar glacier.

After a brief pause, Lauener again strapped on his knapsack, and tempered both will and muscles by the remark,
that our worst piece of work was now before us. From the
place where we sat, the mountain fell precipitously for several hundred feet; and down the weathered crags, and over
the loose shingle which encumbered their ledges, our route
now lay. Lauener was in front, cool and collected, lending
at times a hand to Ramsay, and a word of encouragement
to both of us, while I brought up the rear. I found my
full haversack so inconvenient that I once or twice thought
of sending it down the crags in advance of me, but
Lauener assured me that it would be utterly destroyed
before reaching the bottom. My complaint against it was,
that at critical places it sometimes came between me and
the face of the cliff, pushing me away from the latter so
as to throw my centre of gravity almost beyond the base
intended to support it. We came at length upon a snow-
slope, which had for a time an inclination of 50°; then
once more to the rocks; again to the snow, which was both
steep and deep. Our batons were at least six feet long:
we drove them into the snow to secure an anchorage,
but they sank to their very ends, and we merely retained
a length of them sufficient for a grasp. This slope was
intersected by a so-called Bergschrund, the lower portion
of the slope being torn away from its upper portion so as
to form a crevasse that extended quite round the head of
the valley. We reached its upper edge; the chasm was
partially filled with snow, which brought its edges so near
that we cleared it by a jump. The rest of the slope was

descended by a *glissade*. Each sat down upon the snow,
and the motion, once commenced, swiftly augmented to
the rate of an avalanche, and brought us pleasantly to
the bottom.

As we looked from the heights, we could see that the
valley through which our route lay was filled with gray fog:
into this we soon plunged, and through it we made our way
towards the Abschwung. The inclination of the glacier
was our only guide, for we could see nothing. Reaching
the confluence of the Finsteraar and Lauteraar branches,
we went downwards with long swinging strides, close
alongside the medial moraine of the trunk glacier. The
glory of the morning had its check in the dull gloom of
the evening. Across streams, amid dirt-cones and glacier-
tables, and over the long reach of shingle which covers
the end of the glacier, we plodded doggedly, and reached
the Grimsel at 7 P.M., the journey having cost a little more
than 14 hours.

(15.)

We made the Grimsel our station for a day, which was
spent in examining the evidences of ancient glacier action
in the valley of Hasli. Near the Hospice, but at the
opposite side of the Aar, rises a mountain-wall of hard
granite, on which the flutings and groovings are magni-
ficently preserved. After a little practice the eye can
trace with the utmost precision the line which marks
the level of the ancient ice: above this the crags are
sharp and rugged; while below it the mighty grinder has
rubbed off the pinnacles of the rocks and worn their edges
away. The height to which this action extends must be
nearly two thousand feet above the bed of the present
valley. It is also easy to see the depth to which the river

has worked its channel into the ancient rocks. In some cases the road from Guttanen to the Grimsel lay right over the polished rocks, asperities being supplied by the chisel of man in order to prevent travellers from slipping on their slopes. Here and there also huge protuberant crags were rounded into domes almost as perfect as if chiselled by art. To both my companion and myself this walk was full of instruction and delight.

On the 28th of July we crossed the Grimsel pass, and traced the scratchings to the very top of it. Ramsay remarked that their direction changed high up the pass, as if a tributary from the summit had produced them, while lower down they merged into the general direction of the glacier which had filled the principal valley. From the summit of the Mayenwand we had a clear view of the glacier of the Rhone; and to see the lower portion of this glacier to advantage no better position can be chosen. The dislocation of its cascade, the spreading out of the ice below, its system of radial crevasses, and the transverse sweep of its structural groovings, may all be seen. A few hours afterwards we were amid the wild chasms at the brow of the ice-fall, where we worked our way to the centre of the ice, but were unable to attain the opposite side.

Having examined the glacier both above and below the cascade, we went down the valley to Viesch, and ascended thence, on the 30th of July, to the Hôtel Jungfrau on the slopes of the Æggischhorn. On the following day we climbed to the summit of the mountain, and from a sheltered nook enjoyed the glorious prospect which it commands. The wind was strong, and fleecy clouds flew over the heavens; some of which, as they formed and dispersed themselves about the flanks of the Aletschhorn, showed extraordinary iridescences.

The sunbeams called us early on the morning of the 1st

of August. No cloud rested on the opposite range of the
Valais mountains, but on looking towards the Æggischhorn
we found a cap upon its crest; we looked again—the cap
had disappeared and a serene heaven stretched overhead.
As we breasted the alp the moon was still in the sky,
paling more and more before the advancing day; a single
hawk swung in the atmosphere above us; clear streams
babbled from the hills, the louder sounds reposing on a
base of music; while groups of cows with tinkling bells
browsed upon the green alp. Here and there the grass
was dispossessed, and the flanks of the mountain were
covered by the blocks which had been cast down from the
summit. On reaching the plateau at the base of the final
pyramid, we rounded the mountain to the right and came
over the lonely and beautiful Märjelen See. No doubt the
hollow which this lake fills had been scooped out in former
ages by a branch of the Aletsch glacier; but long ago
the blue ice gave place to blue water. The glacier
bounds it at one side by a vertical wall of ice sixty feet
in height: this is incessantly undermined, a roof of crystal
being formed over the water, till at length the projecting
mass, becoming too heavy for its own rigidity, breaks and
tumbles into the lake. Here, attacked by sun and air,
its blue surface is rendered dazzlingly white, and several
icebergs of this kind now floated in the sunlight; the water
was of a glassy smoothness, and in its blue depths each ice
mass doubled itself by reflection.*

The Aletsch is the grandest glacier in the Alps: over it
we now stood, while the bounding mountains poured vast
feeders into the noble stream. The Jungfrau was in
front of us without a cloud, and apparently so near that I
proposed to my guide to try it without further preparation.
He was enthusiastic at first, but caution afterwards got the

---

* A painting of this exquisite lake has been recently executed by Mr.
George Barnard.

better of his courage. At some distance up the glacier the snow-line was distinctly drawn, and from its edge upwards the mighty shoulders of the hills were heavy laden with the still powdery material of the glacier.

Amid blocks and débris we descended to the ice : the portion of it which bounded the lake had been sapped, and a space of a foot existed between ice and water : numerous chasms were formed here, the mass being thus broken, preparatory to being sent adrift upon the lake. We crossed the glacier to its centre, and looking down it the grand peaks of the Mischabel, the noble cone of the Weisshorn, and the dark and stern obelisk of the Matterhorn, formed a splendid picture. Looking upwards, a series of most singularly contorted dirt-bands revealed themselves upon the surface of the ice. I sought to trace them to their origin, but was frustrated by the snow which overspread the upper portion of the glacier. Along this we marched for three hours, and came at length to the junction of the four tributary valleys which pour their frozen streams into the great trunk valley. The glory of the day, and that joy of heart which perfect health confers, may have contributed to produce the impression, but I thought I had never seen anything to rival in magnificence the region in the heart of which we now found ourselves. We climbed the mountain on the right-hand side of the glacier, where, seated amid the riven and weather-worn crags, we fed our souls for hours on the transcendent beauty of the scene.

We afterwards redescended to the glacier, which at this place was intersected by large transverse crevasses, many of which were apparently filled with snow, while over others a thin and treacherous roof was thrown. In some cases the roof had broken away, and revealed rows of icicles of great length and transparency pendent from the edges. We at length turned our faces homewards, and looking down the glacier I saw at a great distance something moving on the ice. I first thought it was a man, though it seemed strange that a man should be there alone. On drawing my guide's attention to it he at once pronounced it to be a chamois, and I with my telescope immediately verified his statement. The creature bounded up the glacier at intervals, and sometimes the vigour of its spring showed that it had projected itself over a crevasse. It approached us sometimes at full gallop: then would stop, look toward us, pipe loudly, and commence its race once more. It evidently made the reciprocal mistake to my own, imagining us to be of its own kith and kin. We sat down upon the ice the better to conceal our forms, and to its whistle our guide whistled in reply. A joyous rush was the creature's first response to the signal; but it afterwards began to doubt, and its pauses became more frequent. Its form at times was extremely graceful, the head erect in the air, its apparent uprightness being augmented by the curvature which threw its horns back. I watched the animal through my glass until I could see the glistening of its eyes; but soon afterwards it made a final pause, assured itself of its error, and flew with the speed of the wind to its refuge in the mountains.

ASCENT OF THE FINSTERAARHORN, 1858.

(16.)

SINCE my arrival at the hotel on the 30th of July I had once or twice spoken about ascending the Finsteraarhorn, and on the 2nd of August my host advised me to avail myself of the promising weather. A guide, named Bennen, was attached to the hotel, a remarkable-looking man, between 30 and 40 years old, of middle stature, but very strongly built. His countenance was frank and firm, while a light of good-nature at times twinkled in his eye. Altogether the man gave me the impression of physical strength, combined with decision of character. The proprietor had spoken to me many times of the strength and courage of this man, winding up his praises of him by the assurance that if I were killed in Bennen's company there would be two lives lost, for that the guide would assuredly sacrifice himself in the effort to save his *Herr*.

He was called, and I asked him whether he would accompany me alone to the top of the Finsteraarhorn. To this he at first objected, urging the possibility of his having to render me assistance, and the great amount of labour which this might entail upon him; but this was overruled by my engaging to follow where he led, without asking him to render me any help whatever. He then agreed to make the trial, stipulating, however, that he should not have much to carry to the cave of the Faulberg, where we were to spend the night. To this I cordially agreed, and sent on blankets, provisions, wood, and hay, by two porters.

My desire, in part, was to make a series of observations at the summit of the mountain, while a similar series was made by Professor Ramsay in the valley of the Rhone, near

Viesch, with a view to ascertaining the permeability of the
lower strata of the atmosphere to the radiant heat of the
sun. During the forenoon of the 2nd I occupied myself
with my instruments, and made the proper arrangements
with Ramsay. I tested a mountain-thermometer which
Mr. Casella had kindly lent me, and found the boiling
point of water on the dining-room table of the hotel to be
199·29° Fahrenheit. At about three o'clock in the afternoon we quitted the hotel, and proceeded leisurely with
our two guides up the slope of the Æggischhorn. We
once caught a sight of the topmost pinnacle of the Finsteraarhorn; beside it was the Rothhorn, and near this
again the Oberaarhorn, with the Viescher glacier streaming
from its shoulders. On the opposite side we could see,
over an oblique buttress of the mountain on which we
stood, the snowy summit of the Weisshorn; to the left of
this was the ever grim and lonely Matterhorn; and farther
to the left, with its numerous snow-cones, each with
its attendant shadow, rose the mighty Mischabel. We
descended, and crossed the stream which flows from the
Märjelen See, into which a large mass of the glacier had
recently fallen, and was now afloat as an iceberg. We passed
along the margin of the lake, and at the junction of water
and ice I bade Ramsay good bye. At the commencement
of our journey upon the ice, whenever we crossed a crevasse, I noticed Bennen watching me; his vigilance, however, soon diminished, whence I gathered that he finally
concluded that I was able to take care of myself. Clouds
hovered in the atmosphere throughout the whole time of
our ascent; one smoky-looking mass marred the glory of
the sunset, but at some distance was another which exhibited colours almost as rich and varied as those of the
solar spectrum. I took the glorious banner thus unfurled
as a sign of hope, to check the despondency which its
gloomy neighbour was calculated to produce.

Two hours' walking brought us near our place of rest; the porters had already reached it, and were now returning. We deviated to the right, and, having crossed some ice-ravines, reached the lateral moraine of the glacier, and picked our way between it and the adjacent mountain-wall. We then reached a kind of amphitheatre, crossed it, and climbing the opposite slope, came to a triple grotto formed by clefts in the mountain. In one of these a pine-fire was soon blazing briskly, and casting its red light upon the surrounding objects, though but half dispelling the gloom from the deeper portions of the cell. I left the grotto, and climbed the rocks above it to look at the heavens. The sun had quitted our firmament, but still tinted the clouds with red and purple; while one peak of snow in particular glowed like fire, so vivid was its illumination. During our journey upwards the Jungfrau never once showed her head, but, as if in ill temper, had wrapped her vapoury veil around her. She now looked more good-humoured, but still she did not quite remove her hood; though all the other summits, without a trace of cloud to mask their beautiful forms, pointed heavenward. The calmness was perfect; no sound of living creature, no whisper of a breeze, no gurgle of water, no rustle of débris, to break the deep and solemn silence. Surely, if beauty be an object of worship, those glorious mountains, with rounded shoulders of the purest white—snow-crested and star-gemmed—were well calculated to excite sentiments of adoration.

I returned to the grotto, where supper was prepared and waiting for me. The boiling-point of water, at the level of the "kitchen" floor, I found to be 196° Fahr. Nothing could be more picturesque than the aspect of the cave before we went to rest. The fire was gleaming ruddily. I sat upon a stone bench beside it, while Bennen was in front with the red light glimmering fitfully over him. My

boiling-water apparatus, which had just been used, was in
the foreground; and telescopes, opera-glasses, haversacks,
wine-keg, bottles, and mattocks, lay confusedly around.
The heavens continued to grow clearer, the thin clouds,
which had partially overspread the sky, melting gradually
away. The grotto was comfortable; the hay sufficient
materially to modify the hardness of the rock, and my position at least sheltered and warm. One possibility remained that might prevent me from sleeping—the snoring
of my companion; he assured me, however, that he did
not snore, and we lay down side by side. The good fellow
took care that I should not be chilled; he gave me the
best place, by far the best part of the clothes, and may have
suffered himself in consequence; but, happily for him,
he was soon oblivious of this. Physiologists, I believe, have
discovered that it is chiefly during sleep that the muscles
are repaired; and ere long the sound I dreaded announced
to me at once the repair of Bennen's muscles and the
doom of my own. The hollow cave resounded to the deep-
drawn snore. I once or twice stirred the sleeper, breaking
thereby the continuity of the phenomenon; but it instantly pieced itself together again, and went on as before.
I had not the heart to wake him, for I knew that upon
him would devolve the chief labour of the coming day.
At half-past one he rose and prepared coffee, and at two
o'clock I was engaged upon the beverage. We afterwards
packed up our provisions and instruments. Bennen bore
the former, I the latter, and at three o'clock we set out.

We first descended a steep slope to the glacier, along
which we walked for a time. A spur of the Faulberg jutted
out between us and the ice-laden valley through which we
must pass; this we crossed in order to shorten our way
and to avoid crevasses. Loose shingle and boulders over-
laid the mountain; and here and there walls of rock
opposed our progress, and rendered the route far from

agreeable. We then descended to the Grünhorn tributary, which joins the trunk glacier at nearly a right angle, being terminated by a saddle which stretches across from mountain to mountain, with a curvature as graceful and as perfect as if drawn by the instrument of a mathematician. The unclouded moon was shining, and the Jungfrau was before us so pure and beautiful, that the thought of visiting the "Maiden" without further preparation occurred to me. I turned to Bennen, and said," Shall we try the Jungfrau?" I think he liked the idea well enough, though he cautiously avoided incurring any responsibility. "If you desire it, I am ready," was his reply. He had never made the ascent, and nobody knew anything of the state of the snow this year; but Lauener had examined it through a telescope on the previous day, and pronounced it dangerous. In every ascent of the mountain hitherto made, ladders had been found indispensable, but we had none. I questioned Bennen as to what he thought of the probabilities, and tried to extract some direct encouragement from him; but he said that the decision rested altogether with myself, and it was his business to endeavour to carry out that decision. "We will attempt it, then," I said, and for some time we actually walked towards the Jungfrau. A gray cloud drew itself across her summit, and clung there. I asked myself why I deviated from my original intention? The Finsteraarhorn was higher, and therefore better suited for the contemplated observations. I could in no wise justify the change, and finally expressed my scruples. A moment's further conversation caused us to "right about," and front the saddle of the Grünhorn.

The dawn advanced. The eastern sky became illuminated and warm, and high in the air across the ridge in front of us stretched a tongue of cloud like a red flame, and equally fervid in its hue. Looking across the trunk glacier, a valley which is terminated by the Lötsch saddle was seen in a straight line with our route, and I often turned to look along this magnificent corridor. The mightiest mountains in the Oberland form its sides; still, the impression which it makes is not that of vastness or sublimity, but of loveliness not to be described. The sun had not yet smitten the snows of the bounding mountains, but the saddle carved out a segment of the heavens which formed a background of unspeakable beauty. Over the rim of the saddle the sky was deep orange, passing upwards through amber, yellow, and vague ethereal green to the ordinary firmamental blue. Right above the snow-curve purple clouds hung perfectly motionless, giving depth to the spaces between them. There was something saintly in the scene. Anything more exquisite I had never beheld.

We marched upwards over the smooth crisp snow to the crest of the saddle, and here I turned to take a last look along that grand corridor, and at that wonderful " daffodil sky." The sun's rays had already smitten the snows of the Aletschhorn; the radiance seemed to infuse a principle of life and activity into the mountains and glaciers, but still that holy light shone forth, and those motionless clouds floated beyond, reminding one of that eastern religion whose essence is the repression of all action and the substitution for it of immortal calm. The Finsteraarhorn now fronted us; but clouds turbaned the head of the giant, and hid it from our view. The wind, however, being north, inspired us with a strong hope that they would melt as the day advanced. I have hardly seen a finer ice-field than that which now lay before us. Considering the *névé* which supplies it, it appeared to me that the Viescher glacier ought to discharge as much ice as the Aletsch; but this is an error due to the extent of *névé* which is here at once visible: since a glance at the map of this portion

of the Oberland shows at once the great superiority of
the mountain treasury from which the Aletsch Glacier
draws support. Still, the ice-field before us was a most
noble one. The surrounding mountains were of imposing
magnitude, and loaded to their summits with snow. Down
the sides of some of them the half-consolidated mass fell in
a state of wild fracture and confusion. In some cases the
riven masses were twisted and overturned, the ledges bent,
and the detached blocks piled one upon another in heaps;
while in other cases the smooth white mass descended
from crown to base without a wrinkle. The valley
now below us was gorged by the frozen material thus
incessantly poured into it. We crossed it, and reached the
base of the Finsteraarhorn, ascended the mountain a little
way, and at six o'clock paused to lighten our burdens and
to refresh ourselves.

The north wind had freshened, we were in the shade,
and the cold was very keen. Placing a bottle of tea
and a small quantity of provisions in the knapsack, and
a few figs and dried prunes in our pockets, we commenced the ascent. The Finsteraarhorn sends down a
number of cliffy buttresses, separated from each other by
wide couloirs filled with ice and snow. We ascended
one of these buttresses for a time, treading cautiously
among the spiky rocks; afterwards we went along the
snow at the edge of the spine, and then fairly parted
company with the rock, abandoning ourselves to the *névé*
of the couloir. The latter was steep, and the snow was
so firm that steps had to be cut in it. Once I paused
upon a little ledge, which gave me a slight footing, and
took the inclination. The slope formed an angle of 45°
with the horizon; and across it, at a little distance below me, a gloomy fissure opened its jaws. The sun now
cleared the summits which had before cut off his rays, and
burst upon us with great power, compelling us to resort to

our veils and dark spectacles. Two years before, Bennen had been nearly blinded by inflammation brought on by the glare from the snow, and he now took unusual care in protecting his eyes. The rocks looking more practicable, we again made towards them, and clambered among them till a vertical precipice, which proved impossible of ascent, fronted us. Bennen scanned the obstacle closely as we slowly approached it, and finally descended to the snow, which wound at a steep angle round its base: on this the footing appeared to me to be singularly insecure, but I marched without hesitation or anxiety in the footsteps of my guide.

We ascended the rocks once more, continued along them for some time, and then deviated to the couloir on our left. This snow-slope is much dislocated at its lower portion, and above its precipices and crevasses our route now lay. The snow was smooth, and sufficiently firm and steep to render the cutting of steps necessary. Bennen took the lead: to make each step he swung his mattock once, and his hindmost foot rose exactly at the moment the mattock descended; there was thus a kind of rhythm in his motion, the raising of the foot keeping time to the swing of the implement. In this manner we proceeded till we reached the base of the rocky pyramid which caps the mountain.

One side of the pyramid had been sliced off, thus dropping down almost a sheer precipice for some thousands of feet to the Finsteraar glacier. A wall of rock, about 10 or 15 feet high, runs along the edge of the mountain, and this sheltered us from the north wind, which surged with the sound of waves against the tremendous barrier at the other side. "Our hardest work is now before us," said my guide. Our way lay up the steep and splintered rocks, among which we sought out the spikes which were closely enough wedged to bear our

weight. Each had to trust to himself, and I fulfilled
to the letter my engagement with Bennen to ask no
help. My boiling-water apparatus and telescope were
on my back, much to my annoyance, as the former was
heavy, and sometimes swung awkwardly round as I twisted
myself among the cliffs. Bennen offered to take it, but he
had his own share to carry, and I was resolved to bear
mine. Sometimes the rocks alternated with spaces of
ice and snow, which we were at intervals compelled to
cross; sometimes, when the slope was pure ice and very
steep, we were compelled to retreat to the highest cliffs.
The wall to which I have referred had given way in
some places, and through the gaps thus formed the
wind rushed with a loud, wild, wailing sound. Through
these spaces I could see the entire field of Agassiz's observations; the junction of the Lauteraar and Finsteraar
glaciers at the Abschwung, the medial moraine between
them, on which stood the Hôtel des Neufchâtelois, and the
pavilion built by M. Dolfuss, in which Huxley and myself
had found shelter two years before. Bennen was evidently
anxious to reach the summit, and recommended all observations to be postponed until after our success had been
assured. I agreed to this, and kept close at his heels.
Strong as he was, he sometimes paused, laid his head upon
his mattock, and panted like a chased deer. He complained of fearful thirst, and to quench it we had only my
bottle of tea: this we shared loyally, my guide praising
its virtues, as well he might. Still the summit loomed
above us; still the angry swell of the north wind, beating
against the torn battlements of the mountain, made wild
music. Upward, however, we strained; and at last, on
gaining the crest of a rock, Bennen exclaimed, in a jubilant
voice, "*Die höchste Spitze!*"—the highest point. In a
moment I was at his side, and saw the summit within a
few paces of us. A minute or two placed us upon

the topmost pinnacle, with the blue dome of heaven above us, and a world of mountains, clouds, and glaciers beneath.

A notion is entertained by many of the guides that if you go to sleep at the summit of any of the highest mountains, you will

"Sleep the sleep that knows no waking."

Bennen did not appear to entertain this superstition; and before starting in the morning, I had stipulated for ten minutes' sleep on reaching the summit, as part compensation for the loss of the night's rest. My first act, after casting a glance over the glorious scene beneath us, was to take advantage of this agreement; so I lay down and had five minutes' sleep, from which I rose refreshed and brisk. The sun at first beat down upon us with intense force, and I exposed my thermometers; but thin veils of vapour soon drew themselves before the sun, and denser mists spread over the valley of the Rhone, thus destroying all possibility of concert between Ramsay and myself. I turned therefore to my boiling-water apparatus, filled it with snow, melted the first charge, put more in, and boiled it; ascertaining the boiling point to be 187° Fahrenheit. On a sheltered ledge, about two or three yards south of the highest point, I placed a minimum-thermometer, in the hope that it would enable us in future years to record the lowest winter temperatures at the summit of the mountain.*

---

* The following note describes the single observation made with this thermometer. Mr. B. informs me that on finding the instrument Bennen swung it in triumph round his head. I fear, therefore, that the observation gives us no certain information regarding the minimum winter-temperature.

"St. Nicholas, 1859, Aug. 25.

"Sir,—On Tuesday last (the 23rd inst.) a party, consisting of Messrs. B., H., R. L., and myself, succeeded in reaching the summit of the Finster Aarhorn under the guidance of Bennen and Melchior André.

It is difficult to convey any just impression of the scene
from the summit of the Finsteraarhorn : one might, it is
true, arrange the visible mountains in a list, stating their
heights and distances, and leaving the imagination to furnish them with peaks and pinnacles, to build the precipices,
polish the snow, rend the glaciers, and cap the highest
summits with appropriate clouds. But if imagination did
its best in this way, it would hardly exceed the reality,
and would certainly omit many details which contribute
to the grandeur of the scene itself. The various shapes
of the mountains, some grand, some beautiful, bathed
in yellow sunshine, or lying black and riven under the
frown of impervious cumuli ; the pure white peaks, cornices, bosses, and amphitheatres; the blue ice rifts, the

We made it an especial object to observe and reset the minimum-
thermometer which you left there last year. On reaching the summit,
before I had time to stop him, Bennen produced the instrument, and it
is just possible that in moving it he may have altered the position
of the index. However, as he held the instrument horizontally, and did
not, as far as I saw, give it any sensible jerk, I have great confidence that
the index remained unmoved.

"The reading of the index was −32° Cent.

"A portion of the spirit extending over about 10½° (and standing between 33° and 43½°) was separated from the rest, but there appeared to
be no data for determining when the separation had taken place. As it
appeared desirable to unite the two portions of spirit before again setting
the index to record the cold of another winter, we endeavoured to effect
this by heating the bulb, but unfortunately, just as we were expecting to
see them coalesce, the bulb burst, and I have now to express my great
regret that my clumsiness or ignorance of the proper mode of setting the
instrument in order should have interfered with the continuance of obser
vations of so much interest. The remains of the instrument, together
with a note of the accident, I have left in the charge of Wellig, the land-
lord of the hotel on the Æggischhorn.

"We reached the summit about 10·40 A.M. and remained there till noon ;
the reading of a pocket thermometer in the shade was 41° F.

"Should there be any further details connected with our ascent on which
you would like to have information, I shall be happy to supply them to
the best of my recollection. Meanwhile, with a farther apology for my
clumsiness, I beg to subscribe myself yours respectfully,

                                                    " H."

" Professor Tyndall."

stratified snow-precipices, the glaciers issuing from the hollows of the eternal hills, and stretching like frozen serpents through the sinuous valleys; the lower cloud field—itself an empire of vaporous hills—shining with dazzling whiteness, while here and there grim summits, brown by nature, and black by contrast, pierce through it like volcanic islands through a shining sea,—add to this the consciousness of one's position which clings to one *unconsciously*, that undercurrent of emotion which surrounds the question of one's personal safety, at a height of more than 14,000 feet above the sea, and which is increased by the weird strange sound of the wind surging with the full deep boom of the distant sea against the precipice behind, or rising to higher cadences as it forces itself through the crannies of the weatherworn rocks, — all conspire to render the scene from the Finsteraarhorn worthy of the monarch of the Bernese Alps.

My guide at length warned me that we must be moving; repeating the warning more impressively before I attended to it. We packed up, and as we stood beside each other ready to march he asked me whether we should tie ourselves together, at the same time expressing his belief that it was unnecessary. Up to this time we had been separate, and the thought of attaching ourselves had not occurred to me till he mentioned it. I thought it, however, prudent to accept the suggestion, and so we united our destinies by a strong rope. "Now," said Bennen, "have no fear; no matter how you throw yourself, I will hold you." Afterwards, on another perilous summit, I repeated this saying of Bennen's to a strong and active guide, but his observation was that it was a hardy untruth, for that in many places Bennen could not have held me. Nevertheless a daring word strengthens the heart, and, though I felt no trace of that sentiment which Bennen exhorted me to banish, and was determined, as far as in me lay, to give him no opportunity of trying his strength in saving me, I liked the fearless utterance of the man, and, sprang cheerily after him. Our descent was rapid, apparently reckless, amid loose spikes, boulders, and vertical prisms of rock, where a false step would assuredly have been attended with broken bones; but the consciousness of certainty in our movements never forsook us, and proved a source of keen enjoyment. The senses were all awake, the eye clear, the heart strong, the limbs steady, yet flexible, with power of recovery in store, and ready for instant action should the footing give way. Such is the discipline which a perilous ascent imposes.

We finally quitted the crest of rocks, and got fairly upon the snow once more. We first went downwards at a long swinging trot. The sun having melted the crust which we were compelled to cut through in the morning, the leg at each plunge sank deeply into the snow; but this sinking was partly in the direction of the slope of the mountain, and hence assisted our progress. Sometimes the crust was hard enough to enable us to glide upon it for long distances while standing erect; but the end of these *glissades* was always a plunge and tumble in the deeper snow. Once upon a steep hard slope Bennen's footing gave way; he fell, and went down rapidly, pulling me after him. I fell also, but turning quickly, drove the spike of my hatchet into the ice, got good anchorage, and held both fast; my success assuring me that I had improved as a mountaineer since my ascent of Mont Blanc. We tumbled so often in the soft snow, and our clothes and boots were so full of it, that we thought we might as well try the sitting posture in gliding down. We did so, and descended with extraordinary velocity, being checked at intervals by a bodily immersion in the softer and deeper snow. I was usually in front of Bennen, shooting down with the speed of an arrow, and feeling the check of the rope when the rapidity of my motion exceeded my guide's

estimate of what was safe. Sometimes I was behind him,
and darted at intervals with the swiftness of an avalanche
right upon him; sometimes in the same transverse line with
him, with the full length of the rope between us; and here
I found its check unpleasant, as it tended to make me roll
over. My feet were usually in the air, and it was only
necessary to turn them right or left, like the helm of a
boat, to change the direction of motion and avoid a difficulty, while a vigorous dig of leg and hatchet into the
snow was sufficient to check the motion and bring us to
rest. Swiftly, yet cautiously, we glided into the region of
crevasses, where we at last rose, quite wet, and resumed
our walking, until we reached the point where we had left
our wine in the morning, and where I squeezed the water
from my wet clothes, and partially dried them in the sun.

We had left some things at the cave of the Faulberg,
and it was Bennen's first intention to return that way and
take them home with him. Finding, however, that we
could traverse the Viescher glacier almost to the Æggischhorn, I made this our highway homewards. At the place
where we entered it, and for an hour or two afterwards,
the glacier was cut by fissures, for the most part covered
with snow. We had packed up our rope, and Bennen admonished me to tread in his steps. Three or four times he
half disappeared in the concealed fissures, but by clutch-
ing the snow he rescued himself and went on as swiftly as
before. Once my leg sank, and the ring of icicles some
fifty feet below told me that I was in the jaws of a crevasse; my guide turned sharply—it was the only time
that I had seen concern on his countenance :—

" *Gott's Donner ! Sie haben meine Tritte nicht gefolgt.*"

" *Doch !* " was my only reply, and we went on. He
scarcely tried the snow that he crossed, as from its form
and colour he could in most cases judge of its condition.
For a long time we kept at the left-hand side of the

glacier, avoiding the fissures which were now permanently open. We came upon the tracks of a herd of chamois, which had clambered from the glacier up the sides of the Oberaarhorn, and afterwards crossed the glacier to the right-hand side, my guide being perfect master of the ground. His eyes went in advance of his steps, and his judgment was formed before his legs moved. The glacier was deeply fissured, but there was no swerving, no retreating, no turning back to seek more practicable routes; each stride told, and every stroke of the axe was a profitable investment of labour.

We left the glacier for a time, and proceeded along the mountain side, till we came near the end of the Trift glacier, where we let ourselves down an awkward face of rock along the track of a little cascade, and came upon the glacier once more. Here again I had occasion to admire the knowledge and promptness of my guide. The glacier, as is well known, is greatly dislocated, and has once or twice proved a prison to guides and travellers, but Bennen led me through the confusion without a pause. We were sometimes in the middle of the glacier, sometimes on the moraine, and sometimes on the side of the flanking mountain. Towards the end of the day we crossed what seemed to be the consolidated remains of a great avalanche; on this my foot slipped, there was a crevasse at hand, and a sudden effort was necessary to save me from falling into it. In making this effort the spike of my axe turned uppermost, and the palm of my hand came down upon it, thus inflicting a very angry wound. We were soon upon the green alp, having bidden a last farewell to the ice. Another hour's hard walking brought us to our hotel. No one seeing us crossing the alp would have supposed that we had laid such a day's work behind us; the proximity of home gave vigour to our strides, and our progress was much more speedy than it had been on starting in the morning. I was affectionately welcomed by Ramsay, had a warm bath, dined, went to bed, where I lay fast locked in sleep for eight hours, and rose next morning as fresh and vigorous as if I had never scaled the Finster-aarhorn.

―――――

(17.)

On the 6th of August there was a long fight between mist and sunshine, each triumphing by turns, till at length the orb gained the victory and cleansed the mountains from every trace of fog. We descended to the Märjelen See, and, wishing to try the floating power of its icebergs, at a place where masses sufficiently large approached near to the shore, I put aside a portion of my clothes, and retaining my boots stepped upon the floating ice. It bore me for a time, and I hoped eventually to be able to paddle myself over the water. On swerving a little, however, from the position in which I first stood, the mass turned over and let me into the lake. I tried a second one, which served me in the same manner; the water was too cold to continue the attempt, and there was also some risk of being unpleasantly ground between the opposing surfaces of the masses of ice. A very large iceberg which had been detached some short time previously from the glacier lay floating at some distance from us. Suddenly a sound like that of a waterfall drew our attention towards it. We saw it roll over with the utmost deliberation, while the water which it carried along with it rushed in cataracts down its sides. Its previous surface was white, its present one was of a lovely blue, the submerged crystal having now come to the air. The summerset of this iceberg produced a commotion all over the lake; the floating masses at its edge clashed together, and a mellow glucking sound, due to the lapping of the undulations against the frozen masses, continued long afterwards.

We subsequently spent several hours upon the glacier; and on this day I noticed for the first time a contemporaneous exhibition of *bedding* and *structure* to which I shall refer at another place. We passed finally to the left bank of the glacier, at some distance below the base of the Æggischhorn, and traced its old moraines at intervals along the flanks of the bounding mountain. At the summit of the ridge we found several fine old *roches moutonnées*, on some of which the scratchings of a glacier long departed were well preserved; and from the direction of the scratchings it might be inferred that the ice moved down the mountain towards the valley of the Rhone. A plunge into a lonely mountain lake ended the day's excursion.

On the 7th of August we quitted this noble station. Sending our guide on to Viesch to take a conveyance and proceed with our luggage down the valley, Ramsay and myself crossed the mountains obliquely, desiring to trace the glacier to its termination. We had no path, but it was hardly possible to go astray. We crossed spurs, climbed and descended pleasant mounds, sometimes with the soft grass under our feet, and sometimes knee-deep in rhododendrons. It took us several hours to reach the end of the glacier, and we then looked down upon it merely. It lay couched like a reptile in a wild gorge, as if it had split the mountain by its frozen snout. We afterwards descended to Mörill, where we met our guide and driver; thence down the valley to Visp; and the following evening saw us lodged at the Monte Rosa hotel in Zermatt.

The boiling-point of water on the table of the *salle à manger*, I found to be 202·58° Fahr.

On the following morning I proceeded without my friend to the Görner glacier. As is well known, the end of this glacier has been steadily advancing for several years, and when I saw it, the meadow in front of it was partly shrivelled up by its irresistible advance. I was informed by my host that within the last sixty years forty-four châlets had been overturned by the glacier, the ground on which they stood being occupied by the ice; at present there are others for which a similar fate seems imminent. In thus advancing the glacier merely takes up ground which belonged to it in former ages, for the rounded rocks which rise out of the adjacent meadow show that it had once passed over them.

I had arranged to meet Ramsay this morning on the road to the Riffelberg. The meeting took place, but I then learned that a minute or two after my departure he had received intelligence of the death of a near relative. Thus was our joint expedition terminated, for he resolved to return at once to England. At my solicitation he accompanied me to the Riffel hotel. We had planned an ascent of Monte Rosa together, but the arrangement thus broke down, and I was consequently thrown upon my own resources. Lauener had never made the ascent, but he nevertheless felt confident that we should accomplish it together.

FIRST ASCENT OF MONTE ROSA, 1858.

(18.)

On Monday, the 9th of August, we reached the Riffel, and, by good fortune, on the evening of the same day, my guide's brother, the well-known Ulrich Lauener, also arrived at the hotel on his return from Monte Rosa. From him we obtained all the information possible respecting the ascent, and he kindly agreed to accompany us a little way the next morning, to put us on the right track. At three A.M. the door of my bedroom opened, and Christian Lauener announced to me that the weather was sufficiently good to justify an attempt. The stars were shining overhead; but Ulrich afterwards drew our attention to some heavy clouds which clung to the mountains on the other side of the valley of the Visp; remarking that the weather *might* continue fair throughout the day, but that these clouds were ominous. At four o'clock we were on our way, by which time a gray stratus cloud had drawn itself across the neck of the Matterhorn, and soon afterwards another of the same nature encircled his waist. We proceeded past the Riffelhorn to the ridge above the Görner glacier, from which Monte Rosa was visible from top to bottom, and where an animated conversation in Swiss patois commenced. Ulrich described the slopes, passes, and precipices, which were to guide us; and Christian demanded explanations, until he was finally able to declare to me that his knowledge was sufficient. We then bade Ulrich good-bye, and went forward. All was clear about Monte Rosa, and the yellow morning light shone brightly upon its uppermost snows. Beside the Queen of the Alps was the huge mass of the Lyskamm, with a saddle stretching from the one to the other; next to the Lyskamm came two white rounded mounds, smooth and pure, the Twins Castor and Pollux, and further to the right again the broad brown flank of the Breithorn. Behind us Mont Cervin gathered the clouds more thickly round him, until finally his grand obelisk was totally hidden. We went along the mountain-side for a time, and then descended to the glacier. The surface was hard frozen, and the ice crunched loudly under our feet. There was a hollowness and volume in the sound which require explanation; and this, I think, is furnished by the remarks of Sir John Herschel on those hollow sounds at the Solfaterra, near Naples, from which travellers have inferred the existence of cavities within the mountain. At the place where these sounds are heard the earth is friable, and, when struck, the concussion is reinforced and lengthened by the partial echos from the surfaces of the fragments. The conditions for a similar effect exist upon the glacier, for the ice is disintegrated to a certain depth, and from the innumerable places of rupture little reverberations are sent, which give a length and hollowness to the sound produced by the crushing of the fragments on the surface.

We looked to the sky at intervals, and once a meteor slid across it, leaving a train of sparks behind. The blue firmament, from which the stars shone down so brightly when we rose, was more and more invaded by clouds, which advanced upon us from our rear, while before us the solemn heights of Monte Rosa were bathed in rich yellow sunlight. As the day advanced the radiance crept down towards the valleys; but still those stealthy clouds advanced like a besieging army, taking deliberate possession of the summits, one after the other, while grey skirmishers moved through the air above us. The play of light and shadow upon Monte Rosa was at times beautiful, bars of gloom and zones of glory shifting and alternating from top to bottom of the mountain.

At five o'clock a grey cloud alighted on the shoulder of
the Lyskamm, which had hitherto been warmed by the
lovely yellow light. Soon afterwards we reached the foot
of Monte Rosa, and passed from the glacier to a slope of
rocks, whose rounded forms and furrowed surfaces showed
that the ice of former ages had moved over them; the
granite was now coated with lichens, and between the
bosses where mould could rest were patches of tender
moss. As we ascended, a peal to the right announced the
descent of an avalanche from the Twins; it came heralded
by clouds of ice-dust, which resembled the sphered masses
of condensed vapour which issue from a locomotive. A
gentle snow-slope brought us to the base of a precipice
of brown rocks, round which we wound; the snow was
in excellent order, and the chasms were so firmly bridged
by the frozen mass that no caution was necessary in
crossing them. Surmounting a weathered cliff to our left,
we paused upon the summit to look upon the scene around
us. The snow gliding insensibly from the mountains, or discharged in avalanches from the precipices which it overhung,
filled the higher valleys with pure white glaciers, which were
rifted and broken here and there, exposing chasms and precipices from which gleamed the delicate blue of the half-
formed ice. Sometimes, however, the *névés* spread over wide
spaces without a rupture or wrinkle to break the smoothness of the superficial snow. The sky was now for the
most part overcast, but through the residual blue spaces
the sun at intervals poured light over the rounded bosses
of the mountain.

At half-past seven o'clock we reached another precipice
of rock, to the left of which our route lay, and here
Lauener proposed to have some refreshment; after which
we went on again. The clouds spread more and more,
leaving at length mere specks and patches of blue
between them. Passing some high peaks, formed by

the dislocation of the ice, we came to a place where the *névé* was rent by crevasses, on the walls of which the stratification due to successive snow-falls was shown with great beauty and definition. Between two of these fissures our way now lay : the wall of one of them was hollowed out longitudinally midway down, thus forming a roof above and a ledge below, and from roof to ledge stretched a railing of cylindrical icicles, as if intended to bolt them together. A cloud now for the first time touched the summit of Monte Rosa, and sought to cling to it, but in a minute it dispersed in shattered fragments, as if dashed to pieces for its presumption. The mountain remained for a time clear and triumphant, but the triumph was short-lived : like suitors that will not be repelled, the dusky vapours came ; repulse after repulse took place, and the sunlight gushed down upon the heights, but it was manifest that the clouds gained ground in the conflict.

Until about a quarter past nine o'clock our work was mere child's play, a pleasant morning stroll along the flanks of the mountain ; but steeper slopes now rose above us, which called for more energy, and more care in the fixing of the feet. Looked at from below, some of these slopes appeared precipitous; but we were too well acquainted with the effect of fore-shortening to let this daunt us. At each step we dug our batons into the deep snow. When first driven in, the batons* *dipped* from us, but were brought, as we walked forward, to the vertical, and finally beyond it at the other side. The snow was thus forced aside, a rubbing of the staff against it, and of the snow-particles against each other, being the consequence. We had thus perpetual rupture and regelation ; while the little sounds consequent upon rupture, reinforced by the partial echos from the surfaces of the granules,

---

* My staff was always the handle of an axe an inch or two longer than an ordinary walking-stick.

were blended together to a note resembling the lowing of cows. Hitherto I had paused at intervals to make notes, or to take an angle; but these operations now ceased, not from want of time, but from pure dislike; for when the eye has to act the part of a sentinel who feels that at any moment the enemy may be upon him; when the body must be balanced with precision, and legs and arms, besides performing actual labour, must be kept in readiness for possible contingencies; above all, when you feel that your safety depends upon yourself alone, and that, if your footing gives way, there is no strong arm behind ready to be thrown between you and destruction; under such circumstances the relish for writing ceases, and you are willing to hand over your impressions to the safe keeping of memory.

From the vast boss which constitutes the lower portion of Monte Rosa cliffy edges run upwards to the summit. Were the snow removed from these we should, I doubt not, see them as toothed or serrated crags, justifying the term "*kamm*," or "comb," applied to such edges by the Germans. Our way now lay along such a kamm, the cliffs of which had, however, caught the snow, and been completely covered by it, forming an edge like the ridge of a house-roof, which sloped steeply upwards. On the Lyskamm side of the edge there was no footing, and, if a human body fell over here, it would probably pass through a vertical space of some thousands of feet, falling or rolling, before coming to rest. On the other side the snow-slope was less steep, but excessively perilous-looking, and intersected by precipices of ice. Dense clouds now enveloped us, and made our position far uglier than if it had been fairly illuminated. The valley below us was one vast cauldron, filled with precipitated vapour, which came seething at times up the sides of the mountain. Sometimes this fog would partially clear away, and the light would gleam

upwards from the dislocated glaciers. My guide continually admonished me to make my footing sure, and to fix
at each step my staff firmly in the consolidated snow. At
one place, for a short steep ascent, the slope became hard
ice, and our position a very ticklish one. We hewed our
steps as we moved upwards, but were soon glad to deviate
from the ice to a position scarcely less awkward. The wind
had so acted upon the snow as to fold it over the edge of
the kamm, thus causing it to form a kind of cornice, which
overhung the precipice on the Lyskamm side of the mountain. This cornice now bore our weight: its snow had
become somewhat firm, but it was yielding enough to
permit the feet to sink in it a little way, and thus secure us
at least against the danger of slipping. Here also at each
step we drove our batons firmly into the snow, availing
ourselves of whatever help they could render. Once,
while thus securing my anchorage, the handle of my
hatchet went right through the cornice on which we stood,
and, on withdrawing it, I could see through the aperture into the cloud-crammed gulf below. We continued
ascending until we reached a rock protruding from the
snow, and here we halted for a few minutes. Lauener
looked upwards through the fog. "According to all description," he observed, "this ought to be the last kamm
of the mountain ; but in this obscurity we can see nothing."
Snow began to fall, and we recommenced our journey,
quitting the rocks and climbing again along the edge.
Another hour brought us to a crest of cliffs, at which,
to our comfort, the kamm appeared to cease, and other
climbing qualities were demanded of us.

On the Lyskamm side, as I have said, rescue would be
out of the question, should the climber go over the edge.
On the other side of the edge rescue seemed possible,
though the slope, as stated already, was most dangerously
steep. I now asked Lauener what he would have done,

supposing my footing to have failed on the latter slope.
He did not seem to like the question, but said that he
should have considered well for a moment and then have
sprung after me; but he exhorted me to drive all such
thoughts away. I laughed at him, and this did more to
set his mind at rest than any formal profession of courage
could have done. We were now among rocks: we climbed
cliffs and descended them, and advanced sometimes with
our feet on narrow ledges, holding tightly on to other ledges
by our fingers; sometimes, cautiously balanced, we moved
along edges of rock with precipices on both sides. Once, in
getting round a crag, Lauener shook a book from his
pocket; it was arrested by a rock about sixty or eighty feet
below us. He wished to regain it, but I offered to supply
its place, if he thought the descent too dangerous. He said
he would make the trial, and parted from me. I thought
it useless to remain idle. A cleft was before me, through
which I must pass; so, pressing my knees and back against
its opposite sides, I gradually worked myself to the top. I
descended the other face of the rock, and then, through
a second ragged fissure, to the summit of another pinnacle.
The highest point of the mountain was now at hand, separated from me merely by a short saddle, carved by weathering out the crest of the mountain. I could hear Lauener
clattering after me, through the rocks behind. I dropped
down upon the saddle, crossed it, climbed the opposite cliff,
and " *die Höchste Spitze* " of Monte Rosa was won.

Lauener joined me immediately, and we mutually congratulated each other on the success of the ascent. The
residue of the bread and meat was produced, and a bottle of
tea was also appealed to. Mixed with a little cognac,
Lauener declared that he had never tasted anything like it.
Snow fell thickly at intervals, and the obscurity was very
great; occasionally this would lighten and permit the sun
to shed a ghastly dilute light upon us through the gleaming

vapour. I put my boiling-water apparatus in order, and fixed it in a corner behind a ledge; the shelter was, however, insufficient, so I placed my hat above the vessel. The boiling-point was 184°·92 Fahr., the ledge on which the instrument stood being 5 feet below the highest point of the mountain.

The ascent from the Riffel hotel occupied us about seven hours, nearly two of which were spent upon the kamm and crest. Neither of us felt in the least degree fatigued; I, indeed, felt so fresh, that had another Monte Rosa been planted on the first, I should have continued the climb without hesitation, and with strong hopes of reaching the top. I experienced no trace of mountain sickness, lassitude, shortness of breath, heart-beat, or headache; nevertheless the summit of Monte Rosa is 15,284 feet high, being less than 500 feet lower than Mont Blanc. It is, I think, perfectly certain, that the rarefaction of the air at this height is not sufficient of itself to produce the symptoms referred to; physical exertion must be superadded.

After a few fitful efforts to dispel the gloom, the sun resigned the dominion to the dense fog and the descending snow, which now prevented our seeing more than 15 or 20 paces in any direction. The temperature of the crags at the summit, which had been shone upon by the unclouded sun during the earlier portion of the day, was 60° Fahr.; hence the snow melted instantly wherever it came in contact with the rock. But some of it fell upon my felt hat, which had been placed to shelter the boiling-water apparatus, and this presented the most remarkable and beautiful appearance. The fall of snow was in fact a shower of frozen flowers. All of them were six-leaved; some of the leaves threw out lateral ribs like ferns, some were rounded, others arrowy and serrated, some were close, others reticulated, but there was no deviation from the six-

leaved type. Nature seemed determined to make us
some compensation for the loss of all prospect, and thus
showered down upon us those lovely blossoms of the frost;
and had a spirit of the mountain inquired my choice, the
view, or the frozen flowers, I should have hesitated
before giving up that exquisite vegetation. It was wonderful to think of, as well as beautiful to behold. Let us
imagine the eye gifted with a microscopic power sufficient
to enable it to see the molecules which composed these
starry crystals; to observe the solid nucleus formed and
floating in the air; to see it drawing towards it its allied
atoms, and these arranging themselves as if they moved to
music, and ended by rendering that music concrete. Surely
such an exhibition of power, such an apparent demonstration of a resident intelligence in what we are accustomed
to call "brute matter," would appear perfectly miraculous. And yet the reality would, if we could see it, transcend the fancy. If the Houses of Parliament were built
up by the forces resident in their own bricks and lithologic
blocks, and without the aid of hodman or mason, there
would be nothing intrinsically more wonderful in the process than in the molecular architecture which delighted us
upon the summit of Monte Rosa.

Twice or thrice had my guide warned me that we must
think of descending, for the snow continued to fall heavily,
and the loss of our track would be attended with imminent
peril. We therefore packed up, and clambered downward
among the crags of the summit. We soon left these behind us, and as we stood once more upon the kamm, looking into the gloom beneath, an avalanche let loose from
the side of an adjacent mountain shook the air with its
thunder. We could not see it, could form no estimate of
its distance, could only hear its roar, which coming to us
through the darkness, had an undefinable element of
horror in it. Lauener remarked, "I never hear those

things without a shudder; the memory of my brother comes back to me at the same time." His brother, who was the best climber in the Oberland, had been literally broken to fragments by an avalanche on the slopes of the Jungfrau.

We had been separate coming up, each having trusted to himself, but the descent was more perilous, because it is more difficult to fix the heel of the boot than the toe securely in the ice. Lauener was furnished with a rope, which he now tied round my waist, and forming a noose at the other end, he slipped it over his arm. This to me was a new mode of attachment. Hitherto my guides in dangerous places had tied the ropes round *their* waists also. Simond had done it on Mont Blanc, and Bennen on the Finsteraarhorn, proving thus their willingness to share my fate whatever that might be. But here Lauener had the power of sending me adrift at any moment, should his own life be imperilled. I told him that his mode of attachment was new to me, but he assured me that it would give him more power in case of accident. I did not see this at the time; but neither did I insist on his attaching himself in the usual way. It could neither be called anger nor pride, but a warm flush ran through me as I remarked, that I should take good care not to test his power of holding me. I believe I wronged my guide by the supposition that he made the arrangement with reference to his own safety, for all I saw of him afterwards proved that he would at any time have risked his life to save mine. The flush however did me good, by displacing every trace of anxiety, and the rope, I confess, was also a source of some comfort to me. We descended the kamm, I going first. " Secure your footing before you move," was my guide's constant exhortation, " and make your staff firm at each step." We were sometimes quite close upon the rim of the kamm on the Lyskamm side, and we also followed the depressions which marked our track along the cornice. This I now tried intentionally, and drove the handle of my axe through it once or twice. At two places in descending we were upon the solid ice, and these were some of the steepest portions of the kamm. They were undoubtedly perilous, and the utmost caution was necessary in fixing the staff and securing the footing. These however once past, we felt that the chief danger was over. We reached the termination of the edge, and although the snow continued to fall heavily, and obscure everything, we knew that our progress afterwards was se- cure. There was pleasure in this feeling; it was an agreeable variation of that grim mental tension to which I had been previously wound up, but which in itself was by no means disagreeable.

I have already noticed the colour of the fresh snow upon the summit of the Stelvio pass. Since I observed it there it has been my custom to pay some attention to this point at all great elevations. This morning, as I ascended Monte Rosa, I often examined the holes made in the snow by our batons, but the light which issued from them was scarcely perceptibly blue. Now, however, a deep layer of fresh snow overspread the mountain, and the effect was magnifi- cent. Along the kamm I was continually surprised and delighted by the blue gleams which issued from the broken or perforated stratum of new snow; each hole made by the staff was filled with a light as pure, and nearly as deep, as that of the unclouded firmament. When we reached the bottom of the kamm, Lauener came to the front, and tramped before me. As his feet rose out of the snow, and shook the latter off in fragments, sudden and wonderful gleams of blue light flashed from them. Doubtless the blue of the sky has much to do with mountain colouring, but in the present instance not only was there no blue sky, but the air was so thick with fog and descending snow-

flakes, that we could not see twenty yards in advance of
us. A thick fog, which wrapped the mountain quite
closely, now added its gloom to the obscurity caused by
the falling snow. Before we reached the base of the
mountain the fog became thin, and the sun shone through
it. There was not a breath of air stirring, and, though
we stood ankle-deep in snow, the heat surpassed anything
of the kind I had ever felt: it was the dead suffocating
warmth of the interior of an oven, which encompassed us
on all sides, and from which there seemed no escape.
Our own motion through the air, however, cooled us considerably. We found the snow-bridges softer than in the
morning, and consequently needing more caution; but
we encountered no real difficulty among them. Indeed
it is amusing to observe the indifference with which a
snow-roof is often broken through, and a traveller immersed to the waist in the jaws of a fissure. The effort
at recovery is instantaneous; half instinctively hands and
knees are driven into the snow, and rescue is immediate.
Fair glacier work was now before us; after which we reached
the opposite mountain-slope, which we ascended, and then
went down the flank of the Riffelberg to our hotel.
The excursion occupied us eleven and a half hours.

(19.)

On the afternoon of the 11th I made an attempt alone
to ascend the Riffelhorn, and attained a considerable height;
but I attacked it from the wrong side, and the fading
light forced me to retreat. I found some agreeable people
at the hotel on my return. One clergyman especially,
with a clear complexion, good digestion, and bad lungs—
of free, hearty, and genial manner—made himself extremely pleasant to us all. He appeared to bubble over

with enjoyment, and with him and others on the morning
of the 13th I walked to the Görner Grat, as it lay on
the way to my work. We had a glorious prospect from
the summit: indeed the assemblage of mountains, snow,
and ice, here within view is perhaps without a rival in the
world.* I shouldered my axe, and saying "good-bye"
moved away from my companions.

"Are you going?" exclaimed the clergyman. "Give
me one grasp of your hand before we part."

This was the signal for a grasp all round; and the
hearty human kindness which thus showed itself contributed that day to make my work pleasant to me.

We proceeded along the ridge of the Rothe Kumm to
a point which commanded a fine view of the glacier. The
ice had been over these heights in ages past, for, although
lichens covered the surfaces of the old rocks, they did not
disguise the grooves and scratchings. The surface of the
glacier was now about a thousand feet below us, and this
it was our desire to attain. To reach it we had to descend
a succession of precipices, which in general were weathered
and rugged, but here and there, where the rock was
durable, were fluted and grooved. Once or twice indeed
we had nothing to cling to but the little ridges thus
formed. We had to squeeze ourselves through narrow
fissures, and often to get round overhanging ledges, where
our main trust was in our feet, but where these had only
ledges an inch or so in width to rest upon. These cases
were to me the most unpleasant of all, for they compelled
the arms to take a position which, if the footing gave way,
would necessitate a *wrench*, for which I entertain considerable abhorrence. We came at length to a gorge by which
the mountain is rent from top to bottom, and into which
we endeavoured to descend. We worked along its rim for

* In 1858 Mr. E. W. Cooke made a pencil-sketch of this splendid
panorama, which is the best and truest that I have yet seen.

a time, but found its smooth faces too deep. We retreated;
Lauener struck into another track, and while he tested
it I sat down near some grass tufts, which flourished on
one of the ledges, and found the temperature to be as
follows:—

Temperature of rock	..	..	42° C.
Of air an inch above the rock	..		32
Of air a foot from rock	..	..	22
Of grass	..	..	..	25

The first of these numbers does not fairly represent the
temperature of the rock, as the thermometer could be in
contact with it only at one side at a time. It was differences such as these between grass and stone, producing a
mixed atmosphere of different densities, that weakened the
sound of the falls of the Orinoco, as observed and explained
by Humboldt.

By a process of " trial and error " we at length reached
the ice, after two hours had been spent in the effort to
disentangle ourselves from the crags. The glacier is
forcibly thrust at this place against the projecting base
of the mountain, and the structure of the ice correspondingly developed. Crevasses also intersect the ice,
and the blue veins cross them at right angles. I ascended the glacier to a region where the ice was compressed and greatly contorted, and thought that in some
cases I could see the veins crossing the lines of stratification. Once my guide drew my attention to what he
called " *ein sonderbares Loch.*" On one of the slopes an
archway was formed which appeared to lead into the body
of the glacier. We entered it, and explored the cavern to
its end. The walls were of transparent blue ice, singularly
free from air-bubbles; but where the roof of the cavern
was thin enough to allow the sun to shine feebly through
it, the transmitted light was of a pink colour. My guide

expressed himself surprised at *das "röthliche Schein."* At one place a plate of ice had been placed like a ceiling across the cavern; but owing to lateral squeezing it had been broken so as to form a V. I found some air-bubbles in this ice, and in all cases they were associated with blebs of water. A portion of the "ceiling," indeed, was very full of bubbles, and was at some places reduced, by internal liquefaction, to a mere skeleton of ice, with water-cells between its walls.

High up the glacier (towards the old Weissthor) the horizontal stratification is everywhere beautifully shown. I drew my guide's attention to it, and he made the remark that the perfection of the lower ice was due to the pressure of the layers above it. "The snow by degrees compressed itself to glacier." As we approached one of the tributaries on the Monte Rosa side, where great pressure came into play, the stratification appeared to yield and the true structure to cross it at those places where it had yielded most. As the place of greatest pressure was approached, the bedding disappeared more and more, and a clear vertical structure was finally revealed.

THE GÖRNER GRAT AND THE RIFFELHORN.
MAGNETIC PHENOMENA.

(20.)

AT an early hour on Saturday, the 14th of August, I heard the servant exclaim, "*Das Wetter ist wunderschön;*" which good news caused me to spring from my bed and prepare to meet the morn. The range of summits at the opposite side of the valley of St. Nicholas was at first quite clear, but as the sun ascended light cumuli formed round them, increasing in density up to a certain point; below these clouds the air of the valley was transparent; above them the air of heaven was still more so; and thus they swung midway between heaven and earth, ranging themselves in a level line along the necks of the mountains.

It might be supposed that the presence of the sun heating the air would tend to keep it more transparent, by increasing its capacity to dissolve all visible cloud; and this indeed is the true action of the sun. But it is not the only action. His rays, as he climbed the eastern heaven, shot more and more deeply into the valley of St. Nicholas, the moisture of which rose as invisible vapour, remaining unseen as long as the air possessed sufficient warmth to keep it in the vaporous state. High up, however, the cold crags which had lost their heat by radiation the night before, acted like condensers upon the ascending vapour, and caused it to curdle into visible fog. The current, however, continued ascensional, and the clouds were slowly lifted above the tallest peaks, where they arranged themselves in fantastic forms, shifting and changing shape as they gradually melted

away. One peak stood like a field-officer with his cap
raised above his head, others sent straggling cloud-balloons
upwards; but on watching these outlyers they were gradually seen to disappear. At first they shone like snow
in the sunlight, but as they became more attenuated they
changed colour, passing through a dull red to a dusky
purple hue, until finally they left no trace of their
existence.

As the day advanced, warming the rocks, the clouds
wholly disappeared, and a hyaline air formed the setting
of both glaciers and mountains. I climbed to the Görner
Grat to obtain a general view of the surrounding scene.
Looking towards the origin of the Görner glacier the view
was bounded by a wide col, upon which stood two lovely
rounded eminences enamelled with snow of perfect purity.
They shone like burnished silver in the sunlight, as if
their surfaces had been melted and recongealed to frosted
mirrors from which the rays were flung. To the right
of these were the bounding crags of Monte Rosa, and
then the body of the mountain itself, with its crest of
crag and coat of snows. To the right of Monte Rosa,
and almost rivalling it in height, was the vast mass
of the Lyskamm, a rough and craggy mountain, to
whose ledges clings the snow which cannot grasp its
steeper walls, sometimes leaning over them in impending
precipices, which often break, and send wild avalanches
into the space below. Between the Lyskamm and Monte
Rosa lies a large wide valley into which both mountains
pour their snows, forming there the Western glacier of
Monte Rosa—a noble ice stream, which from its magnitude and permanence deserves to impose its name
upon the trunk glacier. It extends downwards from the
col which unites the two mountains; riven and broken
at some places, but at others stretching white and pure
down to its snow-line, where the true glacier emerges from

the *névé*. From the rounded shoulders of the Twin Castor
a glacier descends, at first white and shining, then suddenly broken into faults, fissures, and precipices, which
are afterwards repaired, and the glacier joins that of
Monte Rosa before the junction of the latter with the
trunk stream. Next came a boss of rock, with a secondary
glacier clinging to it as if plastered over it, and after it
the Schwarze glacier, bounded on one side by the Breithorn,
and on the other by the Twin Pollux. This glacier is of
considerable magnitude. Over its upper portion rise the
Twin eminences, pure and white; then follows a smooth and
undulating space, after passing which the *névé* is torn
up into a collection of peaks and chasms; these, however, are mended lower down, and the glacier moves
smoothly and calmly to meet its brothers in the main
valley. Next comes the Trifti glacier,* embraced on all
sides by the rocky arms of the Breithorn; its mass is not
very great, but it descends in a graceful sweep, and
exhibits towards its extremity a succession of beautiful
bands. Afterwards we have the glacier of the Petit
Mont Cervin and those of St. Theodule, which latter are
the last that empty their frozen cargos into the valley
of the Görner. All the glaciers here mentioned are
welded together to a common trunk which squeezes itself
through the narrow defile at the base of the Riffelhorn.
Soon afterwards the moraines become confused, the glacier
drops steeply to its termination, and ploughs up the
meadows in front of it with its irresistible share.

In a line with the Riffelhorn, and rising over the
latter so high as to make it almost vanish by comparison,
was the Titan obelisk of the Matterhorn, from the base of
which the Furgge glacier struggles downwards. On the
other side are the Zmutt glacier, the Schönbuhl, and the

---

* I take this name from Studer's map. Sometimes, however, I have
called it the " Breithorn glacier."

Hochwang, from the Dent Blanche; the Gabelhorn and
Trift glaciers, from the summits which bear those names.
Then come the glaciers of the Weisshorn. Describing a
curve still farther to the right we alight on the peaks
of the Mischabel, dark and craggy precipices from this
side, though from the Æggischhorn they appear as cones
of snow. Sweeping by the Alphubel, the Allaleinhorn,
the Rympfischorn, and Strahlhorn—all of them majestic—
we reach the pass of the Weissthor, and the Cima di
Jazzi. This completes the glorious circuit within the
observer's view.

I placed my compass upon a piece of rock to find the
bearing of the Görner glacier, and was startled at seeing
the sun and it at direct variance. What the sun declared
to be north, the needle affirmed to be south. I at first
supposed that the maker had placed the S where the N
ought to be, and *vice versâ*. On shifting my position,
however, the needle shifted also, and I saw immediately
that the effect was due to the rock of the Grat. Sometimes one end of the needle *dipped* forcibly, at other
places it whirled suddenly round, indicating an entire
change of polarity. The rock was evidently to be regarded as an assemblage of magnets, or as a single
magnet full of "consequent points." A distance of transport not exceeding an inch was, in some cases, sufficient
to reverse the position of the needle. I held the needle
between the two sides of a long fissure a foot wide.
The needle set *along* the fissure at some places, while at
others it set *across* it. Sometimes a little jutting knob
would attract the north end of the needle, while a closely
adjacent little knob would forcibly repel it, and attract
the south end. One extremity of a ledge three feet long
was north magnetic, the other end was south magnetic,
while a neutral point existed midway between both, the
ledge having therefore the exact polar arrangement of an

ordinary bar-magnet. At the highest point of the rock the action appeared to be most intense, but I also found an energetic polarity in a mass at some distance below the summit.

Remembering that Professor Forbes had noticed some peculiar magnetic effect upon the Riffelhorn, I resolved to ascend it. Descending from the Grat we mounted the rocks which form the base of the horn; these are soft and soapy from the quantity of mica which they contain; the higher rocks of the horn are, however, very dense and hard. The ascent is a pleasant bit of mountain practice. We climbed the walls of rock, and wound round the ledges, seeking the assailable points. I tried the magnetic condition of the rocks as we ascended, and found it in general feeble. In other respects the Riffelhorn is a most remarkable mass. The ice of the Görner glacier of former ages, which rose hundreds, perhaps thousands of feet above its present level, encountered the horn in its descent, and was split by the latter, a diversion of the ice along the sides of the peak being the consequence. Portions of the vertical walls of the horn are polished by this action as if they had come from the hands of a lapidary, and the scratchings are as sharp and definite as if drawn by points of steel. I never saw scratchings so perfectly preserved: the finest lines are as clear as the deepest, a consequence of the great density and durability of the rock. The latter evidently contains a good deal of iron, and its surface near the summit is of the rich brown red due to the peroxide of the metal. When we fairly got among the precipices we left our hatchets behind us, trusting subsequently to our hands and feet alone. Squeezing, creeping, clinging, and climbing, in due time we found ourselves upon the summit of the horn.

A pile of stones had been erected near the point where we gained the top. I examined the stones of this pile, and found them strongly polar. The surrounding rocks also showed a violent action, the needle oscillating quickly, and sometimes twirling swiftly round upon a slight change of position. The fragments of rock scattered about were also polar. Long ledges shewed north magnetism for a considerable length, and again for an equal length south magnetism. Two parallel masses separated from each other by a fissure, showed the same magnetic distribution. While I was engaged at one end of the horn, Lauener wandered to the other, on which stood two or three *hommes de pierres.* He was about disturbing some of the stones, when a yell from me surprised him. In fact, the thought had occurred to me that the magnetism of the horn had been developed by lightning striking upon it, and my desire was to examine those points which were most exposed to the discharge of the atmospheric electricity; hence my shout to my guide to let the stones alone. I worked towards the other end of the horn, examining the rocks in my way. Two weathered prominences, which seemed very likely recipients of the lightning, acted violently upon the needle. I sometimes descended a little way, and found that among the rocks below the summit the action was greatly enfeebled. On reaching another very prominent point, I found its extremity all north polar, but at a little distance was a cluster of consequent points, among which the transport of a few inches was sufficient to turn the needle round and round.

The piles of stone at the Zermatt end of the horn did not seem so strongly polar as the pile at the other end, which was higher; still a strong polar action was manifested at many points of the surrounding rocks. Having completed the examination of the summit, I descended the horn, and examined its magnetic condition as I went along. It seemed to me that the jutting prominences always exhibited the strongest action. I do not indeed

remember any case in which a strong action did not exhibit itself at the ends of the terraces which constitute
the horn.   In all cases, however, the rock acted as a
number of magnets huddled confusedly together, and not
as if its entire mass was endowed with magnetism of one
kind.

On the evening of the same day I examined the lower
spur of the Riffelhorn.   Amid its fissures and gulleys one
feels as if wandering through the ruins of a vast castle or
fortification; the precipices are so like walls, and the
scratching and polishing so like what might be done by
the hands of man.   I found evidences of strong polar
action in some of the rocks low down.   In the same continuous mass the action would sometimes exhibit itself
over an area of small extent, while the remainder of the
rock showed no appreciable action.   Some of the boulders
cast down from the summit exhibited a strong and varied
polarity.   Fig. 8 is a sketch of one of these; the barbed
end of each arrow repre-
sents the north end of the
needle, which assumed
the various positions
shown in the figure. Mid-
way down the spur I
lighted upon a trans-
verse wall of rock, which
formed in earlier ages

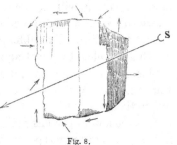

Fig. 8.

the boundary of a lateral outlet of the Görner glacier.   It
was red and hard, weathered rough at some places, and
polished smooth at others.   The lines were drawn finely
upon it, but its outer surface appeared to be peeling off
like a crust; the polished layer rested upon the rock like
a kind of enamel.   The action of the glacier appeared to
resemble that of the break of a locomotive upon rails, both
being cases of exfoliation brought about by pressure and

friction. This wall measured twenty-eight yards across, and one end of it, for a distance of ten or twelve yards, was all north polar; the other end for a similar distance was south polar, but there was a pair of consequent points at its centre.

To meet the case of my young readers, I will here say a few words about the magnetic force. The common magnetic needle points nearly north and south; and if a bit of *iron* be brought near to either end of the needle, they will mutually attract each other. A piece of lead will not show this effect, nor will copper, gold, nor silver. Iron, in fact, is a *magnetic metal*, which the others are not. It is to be particularly observed, that the bit of iron attracts *both ends* of the needle when it is presented to them in succession; and if a common steel sewing needle be substituted for the iron it will be seen that it also has the power of attracting *both ends* of the magnetic needle. But if the needle be rubbed once or twice along one end of a magnet, it will be found that one of its ends will afterwards *repel* a certain end of the magnetic needle and attract the other. By rubbing the needle on the magnet, we thus develop both attraction and repulsion, and this double action of the magnetic force is called its *polarity;* thus the steel which was at first simply *magnetic*, is now magnetic and *polar*.

It is the aim of persons making magnets, that each magnet should have but *two* poles, at its two ends; but it is quite easy to develop in the same piece of steel several pairs of poles; and if the magnetization be irregular, this is sometimes done when we wish to avoid it. These irregular poles are called *consequent points*.

Now I want my young reader to understand that it is not only because the rocks of the Görner Grat and Riffelhorn contain iron, that they exhibit the action which I have described. They are not only *magnetic*, as common

iron is, but, like the magnetized steel needle, they are magnetic and polar. And these poles are irregularly distributed like the " consequent points " to which I have
referred, and this is the reason why I have used the term.

Professor Forbes, as I have already stated, was the first to
notice the effect of the Riffelhorn upon the magnetic needle,
but he seems to have supposed that the entire mass of the
mountain exercised " a local attraction " upon the needle ;
(upon which end he does not say). To enable future observers
to allow for this attraction, he took the bearing of several of
the surrounding mountains from the Riffelhorn ; but it is
very probable that had he changed his position a few inches,
and perfectly certain had he changed it a few yards, he would
have found a set of bearings totally different from those
which he has recorded. The close proximity and irregular
distribution of its consequent points would prevent the
Riffelhorn from exerting any appreciable influence on *a
distant needle,* as in this case the local poles would effectually neutralize each other.

(21.)

On the morning of the 15th the Riffelberg was swathed
in a dense fog, through which heavy rain showered incessantly. Towards one o'clock the continuity of the gray
mass was broken, and sky-gleams of the deepest blue
were seen through its apertures ; these would close up
again, and others open elsewhere, as if the fog were
fighting for existence with the sun behind it. The sun,
however, triumphed, the mountains came more and more
into view, and finally the entire air was swept clear. I
went up to the Görner Grat in the afternoon, and examined
more closely the magnetism of its rocks ; here, as on the
Riffelhorn, I found it most pronounced at the jutting pro-

H

minences of the Grat. Can it be that the superior expo-
sure is more favourable to the formation of the magnetic
oxide of iron? I secured a number of fragments, which I
still possess, and which act forcibly upon a magnetic needle.
The sun was near the western horizon, and I remained
alone upon the Grat to see his last beams illuminate the
mountains, which, with one exception, were without a trace
of cloud. This exception was the Matterhorn, the appearance of which was extremely instructive. The obelisk
appeared to be divided in two halves by a vertical line
drawn from its summit half way down, to the windward of
which we had the bare cliffs of the mountain; and to the
left of it a cloud which appeared to cling tenaciously to the
rocks. In reality, however, there was no clinging; the condensed vapour incessantly got away, but it was ever
renewed, and thus a river of cloud had been sent from
the mountain over the valley of Aosta. The wind in fact
blew lightly up the valley of St. Nicholas charged with
moisture, and when the air that held it rubbed against the
cold cone of the Matterhorn the vapour was chilled and
precipitated in his lee. The summit seemed to smoke
sometimes like a burning mountain; for immediately after
its generation, the fog was drawn away in long filaments
by the wind. As the sun sank lower the ruddiness of his
light augmented, until these filaments resembled streamers
of flame. The sun sank deeper, the light was gradually
withdrawn, and where it had entirely vanished it left the
mountain like a desolate old man whose

> " hoary hair
> Stream'd like a meteor in the troubled air."

For a moment after the sun had disappeared the scene was
amazingly grand. The distant west was ruddy, copious gray
smoke-wreaths were wafted from the mountains, while high
overhead, in an atmospheric region which seemed perfectly
motionless, floated a broad thin cloud, dyed with the richest

iridescences. The colours were of the same character as those which I had seen upon the Aletschorn, being due to interference, and in point of splendour and variety far exceeded anything ever produced by the mere coloured light of the setting sun.

On the 16th I was early upon the glacier. It had frozen hard during the night, and the partially liberated streams flowed, in many cases, over their own ice. I took some clear plates from under the water, and found in them numerous liquid cells, each associated with an air-bubble or a vacuous spot. The most common shape of the cells was a regular hexagon, but there were all forms between the perfect hexagon and the perfect circle. Many cells had also crimped borders, intimating that their primitive form was that of a flower with six leaves. A plate taken from ice which was defended from the sunbeams by the shadow of a rock had no such cells; so that those that I observed were probably due to solar radiation.

My first aim was to examine the structure of the Görnerhorn glacier, which descends the breast of Monte Rosa until it is abruptly cut off by the great Western glacier of the mountain. Between them is a moraine which is at once terminal as regards the former, and lateral as regards the latter. The ice is veined vertically along the moraine, the direction of the structure being parallel to the latter. I ascended the glacier, and found, as I retreated from the place where the thrust was most violent, that the structure became more feeble. From the glacier I passed to the rocks called *auf der Platte*, so as to obtain a general view of its terminal portion. The gradual perfecting of the structure as the region of pressure was approached was very manifest: the ice at the end seemed to wrinkle up in obedience to the pressure, the structural furrows, from being scarcely visible, became more and more decided, and the lamination underneath correspondingly

pronounced, until it finally attained a state of great perfection.

I now quitted the rocks and walked straight across
the Western glacier of Monte Rosa to its centre, where
I found the structure scarcely visible. I next faced the
Görner Grat, and walked down the glacier towards the
moraine which divides it from the Görner glacier. The
mechanical conditions of the ice here are quite evident;
each step brought me to a place of greater pressure, and
also to a place of more highly developed structure, until
finally near to the moraine itself, and running parallel to
it, a magnificent lamination was developed. Here the
superficial groovings could be traced to great distances,
and beside the moraine were boulders poised on pedestals
of ice through which the blue veins ran. At some places
the ice had been weathered into laminæ not more than
a line in thickness.

I now recrossed the Monte Rosa glacier to its junction
with the Schwartze glacier, which descends between the
Twins and Breithorn. The structure of the Monte Rosa
glacier is here far less pronounced than at the other
side, and the pressure which it endures is also manifestly
less; the structure of the Schwartze glacier is fairly
developed, being here parallel to its moraine. The cliffs of
the Breithorn are much exposed to weathering action, and
boulders are copiously showered down upon the adjacent
ice. Between the Schwartze glacier and the glacier
which descends from the breast of the Breithorn itself
these blocks ride upon a spine of ice, and form a moraine
of grand proportions. From it a fine view of the glacier
is attainable, and the gradual development of its structure
as the region of maximum pressure is approached is very
plain. A number of gracefully curved undulations sweep
across the Breithorn glacier, which are squeezed more
closely together as the moraine is approached. All the gla-

ciers that descend from the flanking mountains of the
Görner valley are suddenly turned aside where they meet
the great trunk stream, and are reduced by the pressure
to narrow stripes of ice separated from each other by
parallel moraines.

I ascended the Breithorn glacier to the base of an ice-
fall, on one side of which I found large crumples produced
by the pressure, the veined structure being developed at
right angles to the direction of the latter. No such structure was visible above this place. The crumples were cut
by fissures, perpendicular to which the blue veins ran.
I now quitted the glacier, and clambered up the adjacent
alp, from which a fine view of the general surface was
attainable. As in the case of the Görnerhorn glacier, the
gradual perfecting of the structure was very manifest;
the dirt, which first irregularly scattered over the surface,
gradually assumed a striated appearance, and became
more and more decided as the moraine was approached.
I now descended from the alp, and endeavoured to measure
some of the undulations; proceeding afterwards to the
junction of the Breithorn glacier with that of St. Théodule.
The end of the latter appears to be crumpled by its
thrust against the former, and the moraine between them,
instead of being raised, runs along a hollow which is
flanked by the crumples on either side. The Breithorn
glacier became more and more attenuated, until finally it
actually vanished under its own moraines. On the sides
of the crevasses, by which the Théodule glacier is here
intersected, I thought I could plainly see two systems of
veins cutting each other at an angle of fifteen or twenty
degrees. Reaching the Görner glacier, at a place where
its dislocation was very great, I proceeded down it past
the Riffelhorn, to a point where it seemed possible to scale
the opposite mountain wall. Here I crossed the glacier,
treading with the utmost caution along the combs of ice,

and winding through the entanglement of crevasses until the spur of the Riffelhorn was reached; this I climbed to its summit, and afterwards crossed the Green Alp to our hotel.

The foregoing good day's work was rewarded by a sound sleep at night. The tourists were called in succession next morning, but after each call I instantly subsided into deep slumber, and thus healthily spaced out the interval of darkness. Day at length dawned and gradually brightened. I looked at my watch and found it twenty minutes to six. My guide had been lent to a party of gentlemen who had started at three o'clock for the summit of Monte Rosa, and he had left with me a porter who undertook to conduct me to one of the adjacent glaciers. But as I looked from my window the unspeakable beauty of the morning filled me with a longing to see the world from the top of Monte Rosa. I was in exceedingly good condition—could I not reach the summit alone? Trained and indurated as I had been, I felt that the thing was possible; at all events I could try, without attempting anything which was not clearly within my power.

SECOND ASCENT OF MONTE ROSA, 1858.

(22.)

WHETHER my exercise be mental or bodily, I am always most vigorous when cool. During my student life in Germany, the friends who visited me always complained of the low temperature of my room, and here among the Alps it was no uncommon thing for me to wander over the glaciers from morning till evening in my shirt-sleeves. My object now was to go as light as possible, and hence I left my coat and neckcloth behind me, trusting to the sun and my own motion to make good the calorific waste. After breakfast I poured what remained of my tea into a small glass bottle, an ordinary demi-bouteille, in fact; the waiter then provided me with a ham sandwich, and, with my scrip thus frugally furnished, I thought the heights of Monte Rosa might be won. I had neither brandy nor wine, but I knew the immense amount of mechanical force represented by four ounces of bread and ham, and I therefore feared no failure from lack of nutriment. Indeed, I am inclined to think that both guides and travellers often impair their vigour and render themselves cowardly and apathetic by the incessant "refreshing" which they deem it necessary to indulge in on such occasions.

The guide whom Lauener intended for me was at the door; I passed him and desired him to follow me. This he at first refused to do, as he did not recognise me in my shirt-sleeves; but his companions set him right, and he ran after me. I transferred my scrip to his shoulders, and led the way upward. Once or twice he insinuated that that was not the way to the Schwarze-See, and was probably perplexed by my inattention. From the summit of the

ridge which bounds the Görner glacier the whole grand
panorama revealed itself, and on the higher slopes of Monte
Rosa—so high, indeed, as to put all hope of overtaking
them, or even coming near them, out of the question—a
row of black dots revealed the company which had started
at three o'clock from the hotel. They had made remarkably good use of their time, and I was afterwards informed
that the cause of this was the intense cold, which compelled
them to keep up the proper supply of heat by increased
exertion. I descended swiftly to the glacier, and made for
the base of Monte Rosa, my guide following at some distance behind me. One of the streams, produced by superficial melting, had cut for itself a deep wide channel in the
ice; it was not too wide for a spring, and with the aid of
a run I cleared it and went on. Some minutes afterwards
I could hear the voice of my companion exclaiming, in a
tone of expostulation, "No, no, I won't follow you there."
He however made a circuit, and crossed the stream; I
waited for him at the place where the Monte Rosa glacier
joins the rock, "*auf der Platte*," and helped him down the
ice-slope. At the summit of these rocks I again waited for
him. He approached me with some excitement of manner,
and said that it now appeared plain to him that I intended
to ascend Monte Rosa, but that he would not go with me.
I asked him to accompany me to the summit of the next
cliff, which he agreed to do; and I found him of some service to me. He discovered the faint traces of the party in
advance, and, from his greater experience, could keep them
better in view than I could. We lost them, however, near
the base of the cliff at which we aimed, and I went on,
choosing as nearly as I could remember the route followed
by Lauener and myself a week previous, while my guide
took another route, seeking for the traces. The glacier
here is crevassed, and I was among the fissures some distance in advance of my companion. Fear was manifestly

getting the better of him, and he finally stood still, exclaiming, "No man can pass there." At the same moment I discovered the trace, and drew his attention to it;
he approached me submissively, said that I was quite
right, and declared his willingness to go on. We climbed
the cliff, and discovered the trace in the snow above it.
Here I transferred the scrip and telescope to my own
shoulders, and gave my companion a cheque for five francs.
He returned, and I went on alone.

The sun and heaven were glorious, but the cold
was nevertheless intense, for it had frozen bitterly the
night before. The mountain seemed more noble and
lovely than when I had last ascended it; and as I climbed
the slopes, crossed the shining cols, and rounded the vast
snow-bosses of the mountain, the sense of being alone lent
a new interest to the glorious scene. I followed the track
of those who preceded me, which was that pursued by
Lauener and myself a week previously. Once I deviated
from it to obtain a glimpse of Italy over the saddle which
stretches from Monte Rosa to the Lyskamm. Deep below me was the valley, with its huge and dislocated *névé*,
and the slope on which I hung was just sufficiently steep
to keep the attention aroused without creating anxiety.
I prefer such a slope to one on which the thought of
danger cannot be entertained. I become more weary
upon a dead level, or in walking up such a valley as
that which stretches between Visp and Zermatt, than on
a steep mountain side. The *sense* of weariness is often
no index to the expenditure of muscular force : the
muscles may be charged with force, and, if the nervous
excitant be feeble, the strength lies dormant, and we
are tired without exertion. But the thought of peril
keeps the mind awake, and spurs the muscles into
action; they move with alacrity and freedom, and the
time passes swiftly and pleasantly.

Occupied with my own thoughts as I ascended, I sometimes unconsciously went too quickly, and felt the effects of the exertion. I then slackened my pace, allowing each limb an instant of repose as I drew it out of the snow, and found that in this way walking became rest. This is an illustration of the principle which runs throughout nature—to accomplish physical changes, *time* is necessary. Different positions of the limb require different molecular arrangements; and to pass from one to the other requires time. By lifting the leg slowly and allowing it to fall forward by its own gravity, a man may get on steadily for several hours, while a very slight addition to this pace may speedily exhaust him. Of course the normal pace differs in different persons, but in all the power of endurance may be vastly augmented by the prudent outlay of muscular force.

The sun had long shone down upon me with intense fervour, but I now noticed a strange modification of the light upon the slopes of snow. I looked upwards, and saw a most gorgeous exhibition of interference-colours. A light veil of clouds had drawn itself between me and the sun, and this was flooded with the most brilliant dyes. Orange, red, green, blue—all the hues produced by diffraction were exhibited in the utmost splendour. There seemed a tendency to form circular zones of colour round the sun, but the clouds were not sufficiently uniform to permit of this, and they were consequently broken into spaces, each steeped with the colour due to the condition of the cloud at the place. Three times during my ascent similar veils drew themselves across the sun, and at each passage the splendid phenomena were renewed. As I reached the middle of the mountain an avalanche was let loose from the sides of the Lyskamm; the thunder drew my eyes to the place; I saw the ice move, but it was only the tail of the avalanche; still the volume of sound told me that it was a huge one. Suddenly the front of it appeared from behind a projecting rock, hurling its ice-masses with fury into the valley, and tossing its rounded clouds of ice-dust high into the atmosphere. A wild long-drawn sound, multiplied by echos, now descended from the heights above me. It struck me at first as a note of lamentation, and I thought that possibly one of the party which was now near the summit had gone over the precipice. On listening more attentively I found that the sound shaped itself into an English "hurrah!" I was evidently nearing the party, and on looking upwards I could see them, but still at an immense height above me. The summit still rose before them, and I therefore thought the cheer premature. A precipice of ice was now in front of me, around which I wound to the right, and in a few minutes found myself fairly at the bottom of the Kamm.

I paused here for a moment, and reflected on the work before me. My head was clear, my muscles in perfect condition, and I felt just sufficient fear to render me careful. I faced the Kamm, and went up slowly but surely, and soon heard the cheer which announced the arrival of the party at the summit of the mountain. It was a wild, weird, intermittent sound, swelling or falling as the echos reinforced or enfeebled it. In getting through the rocks which protrude from the snow at the base of the last spur of the mountain, I once had occasion to stoop my head, and, on suddenly raising it, my eyes swam as they rested on the unbroken slope of snow at my left. The sensation was akin to giddiness, but I believe it was chiefly due to the absence of any object upon the snow upon which I could converge the axes of my eyes. Up to this point I had eaten nothing. I now unloosed my scrip, and had two mouthfuls of sandwich and nearly the whole of the tea that remained. I found here that my load, light as it was, impeded me. When fine balancing is necessary, the presence of a very light load, to which one is unaccustomed, may introduce an element of danger, and for this reason I here left the residue of my tea and sandwich behind me. A long long edge was now in front of me, sloping steeply upwards. As I commenced the ascent of this, the foremost of those whose cheer had reached me from the summit some time previously, appeared upon the top of the edge, and the whole party was seen immediately afterwards dangling on the Kamm. We mutually approached each other. Peter Bohren, a well-known Oberland guide, came first, and after him came the gentleman in his immediate charge. Then came other guides with other gentlemen, and last of all my guide, Lauener, with his strong right arm round the youngest of the party. We met where a rock protruded through the snow. The cold smote my naked throat bitterly, so to protect it I borrowed a handkerchief from Lauener, bade my new acquaintances good bye, and proceeded upwards. I was soon at the place where the snow-ridge joins the rocks which constitute the crest of the mountain; through these my way lay, every step I took augmenting my distance from all life, and increasing my sense of solitude. I went up and down the cliffs as before, round ledges, through fissures, along edges of rock, over the last deep and rugged indentation, and up the rocks at its opposite side, to the summit.

A world of clouds and mountains lay beneath me. Switzerland, with its pomp of summits, was clear and grand; Italy was also grand, but more than half obscured. Dark cumulus and dark crag vied in savagery, while at other places white snows and white clouds held equal rivalry. The scooped valleys of Monte Rosa itself were magnificent, all gleaming in the bright sunlight—tossed and torn at intervals, and sending from their rents and walls the magical blue of the ice. Ponderous *névés* lay upon the mountains, apparently motionless, but suggesting motion—sluggish, but indicating irresistible dynamic energy, which moved them slowly to their doom in the warmer valleys below. I thought of my position: it was the first time that a man had stood alone upon that wild peak, and were the imagination let loose amid the surrounding agencies, and permitted to dwell upon the perils which separated the climber from his kind, I dare say curious feelings might have been engendered. But I was prompt to quell all thoughts which might lessen my strength, or interfere with the calm application of it. Once indeed an accident made me shudder. While taking the cork from a bottle which is deposited on the top, and which contains the names of those who have ascended the mountain, my axe slipped out of my hand, and slid some thirty feet away from me. The thought of losing it made my flesh creep, for without it descent would be utterly impossible. I regained it, and looked upon it with an affection which might be bestowed upon a living thing, for it was literally my staff of life under the circumstances. One look more over the cloud-capped mountains of Italy, and I then turned my back upon them, and commenced the descent.

The brown crags seemed to look at me with a kind of friendly recognition, and, with a surer and firmer feeling than I possessed on ascending, I swung myself from crag to crag and from ledge to ledge with a velocity which surprised myself. I reached the summit of the Kamm, and saw the party which I had passed an hour and a half before, emerging from one of the hollows of the mountain; they had escaped from the edge which now lay between them and me. The thought of the possible loss of my axe at the summit was here forcibly revived, for without it I dared not take a single step. My first care was to anchor it firmly in the snow, so as to enable it

to bear at times nearly the whole weight of my body.
In some places, however, the anchor had but a loose
hold; the "cornice" to which I have already referred
became granular, and the handle of the axe went through
it up to the head, still, however, remaining loose. Some
amount of trust had thus to be withdrawn from the
staff and placed in the limbs. A curious mixture of care-
lessness and anxiety sometimes fills the mind on such occasions. I often caught myself humming a verse of a
frivolous song, but this was mechanical, and the sub-
stratum of a man's feelings under such circumstances is
real earnestness. The precipice to my left was a continual
preacher of caution, and the slope to my right was hardly
less impressive. I looked down the former but rarely, and
sometimes descended for a considerable time without
looking beyond my own footsteps. The power of a thought
was illustrated on one of these occasions. I had descended
with extreme slowness and caution for some time, when
looking over the edge of the cornice I saw a row of pointed
rocks at some distance below me. These I felt must
receive me if I slipped over, and I thought how before
reaching them I might so break my fall as to arrive at
them unkilled. This thought enabled me to double my
speed, and as long as the spiky barrier ran parallel to my
track I held my staff in one hand, and contented myself
with a slight pressure upon it.

I came at length to a place where the edge was solid ice,
which rose to the level of the cornice, the latter appearing
as if merely stuck against it. A groove ran between the
ice and snow, and along this groove I marched until the
cornice became unsafe, and I had to betake myself to the
ice. The place was really perilous, but, encouraging
myself by the reflection that it would not last long, I carefully and deliberately hewed steps, causing them to dip a
little inward, so as to afford a purchase for the heel of my

boot, never forsaking one till the next was ready, and never wielding my hatchet until my balance was secured. I was soon at the bottom of the Kamm, fairly out of danger, and, full of glad vigour, I bore swiftly down upon the party in advance of me. It was an easy task to me to fuse myself amongst them as if I had been an old acquaintance, and we joyfully slid, galloped, and rolled together down the residue of the mountain.

The only exception was the young gentleman in Lauener's care. A day or two previously he had, I believe, injured himself in crossing the Gemmi, and long before he reached the summit of Monte Rosa his knee swelled, and he walked with great difficulty. But he persisted in ascending, and Lauener, seeing his great courage, thought it a pity to leave him behind. I have stated that a portion of the Kamm was solid ice. On descending this, Mr. F.'s footing gave way, and he slipped forward. Lauener was forced to accompany him, for the place was too steep and slippery to permit of their motion being checked. Both were on the point of going over the Lyskamm side of the mountain, where they would have indubitably been dashed to pieces. "There was no escape there," said Lauener, in describing the incident to me subsequently, "but I saw a possible rescue at the other side, so I sprang to the right, forcibly swinging my companion round; but in doing so, the baton tripped me up; we both fell, and rolled rapidly over each other down the incline. I knew that some precipices were in advance of us, over which we should have gone, so, releasing myself from my companion, I threw myself in front of him, stopped myself with my axe, and thus placed a barrier before him." After some vain efforts at sliding down the slopes on a baton, in which practice I was fairly beaten by some of my new friends, I attached myself to the invalid, and walked with him and Lauener homewards. Had I gone forward with the foremost of the party, I should have completed the expedition to the summit and back in a little better than nine hours.

I think it right to say one earnest word in connexion with this ascent; and the more so as I believe a notion is growing prevalent that half what is said and written about the dangers of the Alps is mere humbug. No doubt exaggeration is not rare, but I would emphatically warn my readers against acting upon the supposition that it is general. The dangers of Mont Blanc, Monte Rosa, and other mountains, are real, and, if not properly provided against, may be terrible. I have been much accustomed to be alone upon the glaciers, but sometimes, even when a guide was in front of me, I have felt an extreme longing to have a second one behind me. Less than two good ones I think an arduous climber ought not to have; and if climbing without guides were to become habitual, deplorable consequences would assuredly sooner or later ensue.

(23.)

The 18th of August I spent upon the Furgge glacier at the base of Mont Cervin, and what it taught me shall be stated in another place. The evening of this day was signalised by the pleasant acquaintances which it gave me. It was my intention to cross the Weissthor on the morning of the 19th, but thunder, lightning, and heavy rain opposed the project, and with two friends I descended, amid pitiless rain, to Zermatt. Next day I walked by way of Stalden to Saas, where I made the acquaintance of Herr Imseng, the Curé, and on the 21st ascended to the Distel Alp. Near to this place the Allelein glacier pushes its huge terminus right across the valley and dams up the

streams descending from the mountains higher up, thus
giving birth to a dismal lake. At one end of this stands
the Mattmark hotel, which was to be my head-quarters for
a few days.

I reached the place in good company. Near to the
hotel are two magnificent boulders of green serpentine,
which have been lodged there by one of the lateral
glaciers; and two of the ladies desiring to ascend one of
these rocks, a friend and myself helped them to the top.
The thing was accomplished in a very spirited way. Indeed the general contrast, in regard to energy, between
the maidens of the British Isles and those of the Continent
and of America is extraordinary. Surely those who talk
of this country being in its old age overlook the physical
vigour of its sons and daughters. They are strong, but
from a combination of the greatest forces we may obtain
a small resultant, because the forces may act in opposite
directions and partly neutralize each other. Herein, in
fact, lies Britain's weakness; it is strength ill directed;
and is indicative rather of the perversity of young blood
than of the precision of mature years.

Immediately after this achievement I was forsaken by
my friends, and remained the only occupant of the hotel.
A dense gray cloud gradually filled the entire atmosphere,
from which the rain at length began to gush in torrents.
The scene from the windows of the hotel was of the
most dismal character; the rain also came through the
roof, and dripped from the ceiling to the floor. I endeavoured to make a fire, but the air would not let
the smoke of the pine-logs ascend, and the biting of
the hydrocarbons was excruciating to the eyes. On the
whole, the cold was preferable to the smoke. During the
night the rain changed to snow, and on the morning of
the 22nd all the mountains were thickly covered. The
gray delta through which a river of many arms ran

into the Mattmark See was hidden; against some of the windows of the *salle à manger* the snow was also piled, obscuring more than half their light. I had sent my guide to Visp, and two women and myself were the only occupants of the place. It was extremely desolate—I felt, moreover, the chill of Monte Rosa in my throat, and the conditions were not favourable to the cure of a cold.

On the 23rd the Allelein glacier was unfit for work; I therefore ascended to the summit of the Monte Moro, and found the Valais side of the pass in clear sunshine, while impenetrable fog met us on the Italian side. I examined the colour of the freshly fallen snow; it was not an ordinary blue, and was even more transparent than the blue of the firmament. When the snow was broken the light flashed forth; when the staff was dug into the snow and withdrawn, the blue gleam appeared; when the staff lay in a hole, although there might be a sufficient space all round it, the coloured light refused to show itself.

My cough kept me awake on the night of the 23rd, and my cold was worse next day. I went upon the Allelein glacier, but found myself by no means so sure a climber as usual. The best guides find that their powers vary; they are not equally competent on all days. I have heard a celebrated Chamouni guide assert that a man's *morale* is different on different days. The morale in my case had a physical basis, and it probably has so in all. The Allalein glacier, as I have said, crosses the valley and abuts against the opposite mountain; here it is forced to turn aside, and in consequence of the thrust and bending it is crumpled and crevassed. The wall of the Mattmark See is a fine glacier section: looked at from a distance, the ridges and fissures appear arranged like a fan. The structure of the crumpled ice varies from the vertical to the horizontal, and the ridges are sometimes split *along* the planes of structure. The aspect of this portion of the glacier from some of the adjacent heights is exceedingly interesting.

On the morning of the 25th I had two hours' clambering over the mountains before breakfast, and traced the action of ancient glaciers to a great height. The valley of Saas in this respect rivals that of Hasli; the flutings and polishings being on the grandest scale. After breakfast I went to the end of the Allelein glacier, where the Saas Visp river rushes from it: the vault was exceedingly fine, being composed of concentric arches of clear blue ice. I spent several hours here examining the intimate structure of the ice, and found the vacuum disks which I shall describe at another place, of the greatest service to me. As at Rosenlaui and elsewhere, they here taught me that the glacier was composed of an aggregate of small fragments, each of which had a definite plane of crystallization. Where the ice was partially weathered the surfaces of division between the fragments could be traced through the coherent mass, but on crossing these surfaces the direction of the vacuum disks changed, indicating a similar change of the planes of crystallization. The blue veins of the glacier went through its component fragments irrespective of these planes. Sometimes the vacuum disks were parallel to the veins, sometimes across them, sometimes oblique to them.

Several fine masses of ice had fallen from the arch upon its floor, and these were disintegrated to the core. A kick, or a stroke of an axe, sufficed to shake masses almost a cubic yard in size into fragments varying not much on either side of a cubic inch. The veining was finely preserved on the concentric arches of the vault, and some of them apparently exhibited its abolition, or at least confusion, and fresh development by new conditions of pressure. The river being deep and turbulent this day, to reach its opposite side I had to climb the glacier and cross over the crown of its highest arch; this enabled me to get quite in front of the vault, to enter it, and closely inspect those portions where the structure appeared to change. I afterwards ascended the steep moraine which lies between the Allelein and the smaller glacier to the left of it, passing to the latter at intervals to examine its structure. I was at length stopped by the dislocated ice; and from the heights I could count a system of seven dirt-bands, formed by the undulations on the surface of the glacier. On my return to the hotel I found there a number of well-known Alpine men who intended to cross the Adler pass on the following day. Herr Imseng was there: he came to me full of enthusiasm, and asked me whether I would join him in an ascent of the Dom: we might immediately attack it, and he felt sure that we should succeed. The Dom is the highest of the Mischabel peaks, and is one of the grandest of the Alps. I agreed to join the Curé, and with this understanding we parted for the night.

Thursday, 26th August.—A wild stormy morning after a wild and rainy night: the Adler Pass being impassable, the mountaineers returned, and Imseng informed me that the Dom must be abandoned. He gave me the statistics of an avalanche which had fallen in the valley some years before. Within the memory of man Saas had never been touched by an avalanche, but a tradition existed that such a catastrophe had once occurred. On the 14th of March, 1848, at eight o'clock in the morning, the Curé was in his room, when he heard the cracking of pine-branches, and inferred from the sound that an avalanche was descending upon the village. It dashed in the windows of his house and filled his rooms with snow; the sound it produced being sufficient to mask the crashing of the timbers of an adjacent house. Three persons were killed. On the 3rd of April, 1849, heavy snow fell at Saas; the Curé waited until it had attained a depth of four feet, and then retreated to Fée. That night an avalanche descended, and in the line of its rush was a house in which five or six and twenty people had collected for safety: nineteen of them were killed. The Curé afterwards showed me the site of the house, and the direction of the avalanche. It passed through a pine wood; and on expressing my surprise that the trees did not arrest it, he replied that the snow was "quite like dust," and rushed among the trees like so much water. To return from Fée to Saas on the day following he found it necessary to carry two planks. Kneeling upon one of them, he pushed the other forward, and transferred his weight to it, drawing the other after him and repeating the same act. The snow was like flour, and would not otherwise bear his weight. Seeing no prospect of fine weather, I descended to Saas on the afternoon of the 26th. I was the only guest at the hotel; but during the evening I was gratified by the unexpected arrival of my friend Hirst, who was on his way over the Monte Moro to Italy.

For the last five days it had been a struggle between the north wind and the south, each edging the other by turns out of its atmospheric bed, and producing copious precipitation; but now the conflict was decided—the north had prevailed, and an almost unclouded heaven overspread the Alps. The few white fleecy masses that remained were good indications of the swift march of the wind in the upper air. My friend and I resolved to have at least one day's excursion together, and we chose for it the glacier of the Fée. Ascending the mountain by a well-beaten path, we passed a number of "Calvaries" filled with tattered saints and Virgins, and soon came upon the rim of a flattened bowl quite clasped by the mountains. In its centre was the little hamlet of Fée, round which were fresh green pastures, and beyond it the perpetual ice and snow. It was exceedingly picturesque—a scene of human beauty and industry where savagery alone was to be expected. The basin had been scooped by glaciers, and as we paused at its entrance the rounded and fluted rocks were beneath our feet. The Alphubel and the Mischabel raised their crowns to heaven in front of us; the newly fallen snow clung where it could to the precipitous crags of the Mischabel, but on the summits it was the sport of the wind. Sometimes it was borne straight upwards in long vertical striæ; sometimes the fibrous columns swayed to the right, sometimes to the left; sometimes the motion on one of the summits would quite subside; anon the white peak would appear suddenly to shake itself to dust, which it yielded freely to the wind. I could see the wafted snow gradually melt away, and again curdle up into true white cloud by precipitation; this in its turn would be pulled asunder like carded wool, and reduced a second time to transparent vapour.

In the middle of the ice of the Fée stands a green alp, not unlike the Jardin; up this we climbed, halting at intervals upon its grassy knolls to inspect the glacier. I aimed at those places where on à priori grounds I should have thought the production of the veined structure most likely, and reached at length the base of a wall of rock from the edge of which long spears of ice depended. Here my friend halted, while Lauener and myself climbed the precipice, and ascended to the summit of the alp. The snow was deep at many places, and our immersions in unseen holes very frequent. From the peak of the Fée Alp a most glorious view is obtained; in point of grandeur it will bear comparison with any in the Alps, and its seclusion gives it an inexpressible charm. We remained for half an hour upon the warm rock, and then descended. It was our habit to jump from the higher ledges into the deep snow below them, in which we wallowed as if it were flour; but on one of these occasions I lighted on a stone, and the shock produced a curious effect upon my hearing. I appeared suddenly to lose the power of appreciating deep sounds, while the shriller ones were comparatively unimpaired. After I rejoined my friend it required attention on my part to hear him when he spoke to me. This continued until I approached the end of the glacier, when suddenly the babblement of streams, and a world of sounds to which I had been before quite deaf burst in upon me. The deafness was probably due to a strain of the tympanum, such as we can produce artificially, and thus quench low sounds, while shrill ones are scarcely affected.

I was anxious to quit Saas early next morning, but the Curé expressed so strong a wish to show us what he called a *Schauderhaftes Loch*—a terrible hole—which he had himself discovered, that I consented to accompany him. We were joined by his assistant and the priest of Fée. The stream from the Fée glacier has cut a deep channel through the rocks, and along the right-hand bank of the stream we ascended. It was very rough with fallen crags and fallen pines amid which we once or twice lost our way. At length we came to an aperture just sufficient to let a man's body through, and were informed by our conductor that our route lay along the little tunnel: he lay down upon his stomach and squeezed himself through it like a marmot. I followed him; a second tunnel, in which, however, we could stand upright, led into a spacious cavern, formed by the falling together of immense slabs of rock which abutted against each other so as to form a roof. It was the very type of a robber den; and when I remarked this, it was at once proposed to sing a verse from Schiller's play. The young clergyman had a powerful voice—he led and we all chimed in.

> " Ein frohes Leben führen wir,
> Ein Leben voller Wonne.
> Der Wald ist unser Nachtquartier,
> Bei Sturm und Wind hanthieren wir,
> Der Mond ist unsre Sonne."

Herr Imseng wore his black coat; the others had taken theirs off, but they wore their clerical hats, black breeches and stockings. We formed a singular group in a singular place, and the echoed voices mingled strangely with the gusts of the wind and the rush of the river.

Soon after I parted from my friend, and descended the valley to Visp, where I also parted with my guide. He had been with me from the 22nd of July to the 29th of August, and did his duty entirely to my satisfaction. He is an excellent iceman, and is well acquainted both with the glaciers of the Oberland and of the Valais. He is strong and good-humoured, and were I to make another expedition of the kind I don't think that I should take any guide in the Oberland in preference to Christian Lauener.

(24.)

It is a singular fact that as yet we know absolutely nothing of the winter temperature of any one of the high Alpine summits. No doubt it is a sufficient justification of our Alpine men, as regards their climbing, *that they like it.* This plain reason is enough; and no man who ever ascended that "bad eminence" Primrose Hill, or climbed to Hampstead Heath for the sake of a freer horizon, can consistently ask a better. As regards physical science, however, the contributions of our mountaineers have as yet been *nil*, and hence, when we hear of the scientific value of their doings, it is simply amusing to the climbers themselves. I do not fear that I shall offend them in the least by my frankness in stating this. Their pleasure is that of overcoming acknowledged difficulties, and of witnessing natural grandeur. But I would venture to urge that our Alpine men will not find their pleasure lessened by embracing a scientific object in their doings. They

have the strength, the intelligence, and let them add to
these the accuracy which physical science now demands,
and they may contribute work of enduring value. Mr.
Casella will gladly teach them the use of his minimum-
thermometers; and I trust that the next seven years
will not pass without making us acquainted with the
winter temperature of every mountain of note in Switzerland.*

I had thought of this subject since I first read the con-
jectures of De Saussure on the temperature of Mont
Blanc; but in 1857 I met Auguste Balmat at the Jardin,
and there learned from him that he entertained the idea
of placing a self-registering thermometer at the summit of the mountain. Balmat was personally a stranger
to me at the time, but Professor Forbes's writings had
inspired me with a respect for him, which this un-
prompted idea of his augmented. He had procured a
thermometer, the graduation of which, however, he feared
was not low enough. As an encouragement to Balmat,
and with the view of making his laudable intentions known,
I communicated them to the Royal Society, and obtained
from the Council a small grant of money to purchase thermometers and to assist in the expenses of an ascent.
I had now the thermometers in my possession; and
having completed my work at Zermatt and Saas, my next
desire was to reach Chamouni and place the instruments
on the top of Mont Blanc. I accordingly descended the
valley of the Rhone to Martigny, crossed the Tête
Noire, and arrived at Chamouni on the 29th of August,
1858.

Balmat was engaged at this time as the guide of Mr.
Alfred Wills, who, however, kindly offered to place him
at my disposal; and also expressed a desire to accom-

* I find with pleasure that my friend Mr John Ball is now exerting
himself in this direction.

I

pany me himself and assist me in my observations. I
gladly accepted a proposal which gave me for companion so determined a climber and so estimable
a man. But Chamouni was rife with difficulties. In
1857 the Guide Chef had the good sense to give me
considerable liberty of action. Now his mood was entirely
changed: he had been "molested" for giving me so much
freedom. I wished to have a boy to carry a small instrument for me up the Mer de Glace—he would not allow it;
I must take a guide. If I ascended Mont Blanc he declared that I must take four guides; that, in short, I must
in all respects conform to the rules made for ordinary
tourists. I endeavoured to explain to him the advantages
which Chamouni had derived from the labours of men of
science; it was such men who had discovered it when it
was unknown, and it was by their writings that the attention of the general public had been called towards it. It
was a bad recompence, I urged, to treat a man of science
as he was treating me. This was urged in vain; he
shrugged his shoulders, was very sorry, but the thing could
not be changed. I then requested to know his superior,
that I might apply to him; he informed me that there
were a President and Commission of guides at Chamouni,
who were the proper persons to decide the question, and
he proposed to call them together on the 31st of August,
at seven P.M., on condition that I was to be present to state
my own case. To this I agreed.

I spent that day quite alone upon the Mer de Glace,
and climbed amid a heavy snow-storm to the Cleft station
over Trelaporte. When I reached the Montanvert I was
wet and weary, and would have spent the night there were
it not for my engagement with the Guide Chef. I descended amid the rain, and at the appointed hour went to
his bureau. He met me with a polite sympathetic shrug;
explained to me that he had spoken to the Commission,

but that it could not assemble *pour une chose comme ça ;* that the rules were fixed, and I must abide by them. "Well," I responded, "you think you have done your duty; it is now my turn to perform mine. If no other means are available I will have this transaction communicated to the Sardinian Government, and I don't think that it will ratify what you have done." The Guide Chef evidently did not believe a word of it.

Previous to taking any further step I thought it right to see the President of the Commission of Guides, who was also Syndic of the commune. I called upon him on the morning of the 1st of September, and, assuming that he knew all about the transaction, spoke to him accordingly. He listened to me for a time, but did not seem to understand me, which I ascribed partly to my defective French pronunciation. I expressed a hope that he did comprehend me; he said he understood my words very well, but did not know their purport. In fact he had not heard a single word about me or my request. He stated with some indignation that, so far from its being a subject on which the Commission could not assemble, it was one which it was their especial duty to take into consideration. Our conference ended with the arrangement that I was to write him an official letter stating the case, which he was to forward to the Intendant of the province of Faucigny resident at Bonneville. All this was done.

I subsequently memorialised the Intendant himself; and Balmat visited him to secure his permission to accompany me. I have to record, that from first to last the Intendant gave me his sympathy and support. He could not alter laws, but he deprecated a "judaical" interpretation of them. His final letter to myself was as follows :—

"Intendance Royale de la Province de Faucigny,
" Monsieur,—   "Bonneville, 11 Septembre, 1858.

"J'apprends avec une véritable peine les diffi-
cultés que vous rencontrez de la part de M. le Guide Chef
pour l'effectuation de votre périlleuse entreprise scienti-
fique, mais je dois vous dire aussi avec regret que ces diffi-
cultés resident dans un réglement fait en vue de la sécurité
des voyageurs, quelque puisse être le but de leurs excursions.

"Désireux néanmoins de vous être utile notamment en
la circonstance, j'invite aujourd'hui même M. le Guide
Chef à avoir égard à votre projet, à faire en sa faveur une
exception au réglement ci devant eu, tant qu'il n'y aura
aucun danger pour votre sûreté et celles des personnes qui
vous accompagneront, et enfin de se prêter dans les limites
de ses moyens et attributions pour l'heureux succès de
l'expédition, dont les conséquences et résultats n'intéressent
pas seulement la science, mais encore la vallée de Cha-
mounix en particulier.

"Agréez, Monsieur,
" l'assurance de ma considération très-distinguée.

"Pour l'Intendant en congé,

"Le Secrétaire,

"DELEGLISE."

While waiting for this permission I employed myself in
various ways. On the 2nd of September I ascended the
Brévent, from which Mont Blanc is seen to great advantage. From Chamouni its vast slopes are so foreshortened
that one gets a very imperfect idea of the extent to be
traversed to reach the summit. What, however, struck
me most on the Brévent was the changed relation of
the Aiguille du Dru and the Aiguille Verte. From Montanvert the former appears a most imposing mass, while

the peak of the latter appears rather dwarfed behind it; but from the Brévent the Aiguille du Dru is a mere pinnacle stuck in the breast of the grander pyramid of the Aiguille Verte.

On the 4th I rose early, and, strapping on my telescope, ascended to the Montanvert, where I engaged a youth to accompany me up the glacier. The heavens were clear and beautiful :—blue over the Aiguille du Dru, blue over the Jorasse and Mont Mallet, deep blue over the pinnacles of Charmoz, and the same splendid tint stretched grandly over the Col du Géant and its Aiguille. No trace of condensation appeared till towards eleven o'clock, when a little black balloon of cloud swung itself over the Aiguilles Rouges. At one o'clock there were two large masses and a little one between them ; while higher up a white veil, almost too thin to be visible, spread over a part of the heavens. At the zenith, however, and south, north, and west, the blue seemed to deepen as the day advanced. I visited the ice-wall at the Tacul, which seemed lower than it was last year ; the cascade of le Géant appeared also far less imposing. Only in the early part of summer do we see the ice in its true grandeur : its edges and surfaces are then sharp and clear, but afterwards its nobler masses shrink under the influence of sun and air. The *séracs* now appeared wasted and dirty, and not the sharp angular ice-castles which rose so grandly when I first saw them. Thirteen men had crossed the Col du Géant on the day previous, and left an ample trace behind them. This I followed nearly to the summit of the fall. The condition of the glacier was totally different from that of the opposite side on the previous year. The ice was riven, burrowed, and honeycombed, but the track amid all was easy: a vigorous English maiden might have ascended the fall without much difficulty. My object now was to examine the structure of the fall ; but the ice was

not in a good condition for such an examination : it was
too much broken. Still a definite structure was in many
places to be traced, and some of them apparently showed
structure and bedding at a high angle to each other, but
I could not be certain of it. I paused at every commanding point of view and examined the ice through my opera-
glass ; but the result was inconclusive. I observed that
the terraces which compose the fall do not front the middle
of the glacier, but turn their foreheads rather towards its
eastern side, and the consequence is that the protuberances
lower down, which are the remains of these terraces, are
highest at the same side. Standing at the base of the
Aiguille Noire, and looking downwards where the Glacier
des Périades pushes itself against the Géant, a series of fine
crumples is formed on the former, cut across by crevasses,
on the walls of which a forward and backward dipping of
the blue veins is exhibited. Huge crumples are also
formed by the Glacier du Géant, which are well seen from
a point nearly opposite the lowest lateral moraine of the
Glacier des Périades. In some cases the upper portions
of the crumples had scaled off so as to form arches of
ice—a consequence doubtless of the pressure.

The beauty of some Alpine skies is treacherous ; in fact
the deepest blue often indicates an atmosphere charged
almost to saturation with aqueous vapour. This was the
case on the present occasion. Soon after reaching Chamouni
in the evening, rain commenced and continued with scarcely
any intermission until the afternoon of the 8th. I had
given up all hopes of being able to ascend Mont Blanc;
and hence resolved to place the thermometers in some
more accessible position. On the 9th accordingly, accompanied by Mr. Wills, Balmat, and some other friends, I
ascended to the summit of the Jardin, where we placed
two thermometers : one in the ice, at a depth of three feet
below the surface; another on a ledge of the highest

rock.* The boiling-point of water at this place was 194·6°
Fahr.

Deep snow was upon the Talèfre, and the surrounding
precipices were also heavily laden. Avalanches thundered
incessantly from the Aiguille Verte and the other mountains. Scarcely five minutes on an average intervened
between every two successive peals; and after the direct
shock of each avalanche had died away the air of the basin
continued to be shaken by the echos reflected from its
bounding walls.

The day was far spent before we had completed our
work. All through the weather had been fine, and towards
evening augmented to magnificence. As we descended
the glacier from the Couvercle the sun was just disappearing, and the western heaven glowed with crimson, which
crept gradually up the sky until finally it reached the
zenith itself. Such intensity of colouring is exceedingly
rare in the Alps; and this fact, together with the known
variations in the intensity of the firmamental blue, justify
the conclusion that the colouring must, in a great measure,
be due to some *variable constituent* of the atmosphere. If
the air were competent to produce these magnificent effects
they would be the rule instead of the exception.

No sooner had the thermometers been thus disposed of
than the weather appeared to undergo a permanent change.
On the 10th it was perfectly fine—not the slightest mist
upon Mont Blanc; on the 11th this was also the case.
Balmat still had the old thermometer to which I have
already referred; it might not do to show the minimum
temperature of the air, but it might show the temperature
at a certain depth below the surface. I find in my own
case that the finishing of work has a great moral value:

---

* The minimum temperature of the subsequent winter, as shown by this
thermometer, was − 6° Fahr., or 38° below the freezing point. The instrument placed in the ice was broken.

work completed is a safe fulcrum for the performance of other work; and even though in the course of our labours experience should show us a better means of accomplishing a given end, it is often far preferable to reach the end, even by defective means, than to swerve from our course. The habits which this conviction had superinduced no doubt influenced me when I decided on placing Balmat's thermometer on the summit of Mont Blanc.

SECOND ASCENT OF MONT BLANC, 1858.

(25.)

ON the 12th of September, at 5½ A.M. the sunbeams had already fallen upon the mountain; but though the sky above him, and over the entire range of the Aiguilles, was without a cloud, the atmosphere presented an appearance of turbidity resembling that produced by the dust and thin smoke mechanically suspended in a London atmosphere on a dry summer's day. At 20 minutes past 7 we quitted Chamouni, bearing with us the good wishes of a portion of its inhabitants.

A lady accompanied us on horseback to the point where the path to the Grands Mulets deviates from that to the Plan des Aiguilles; here she turned to the left, and we proceeded slowly upwards, through woods of pine, hung with fantastic lichens: escaping from the gloom of these, we emerged upon slopes of bosky underwood, green hazel, and green larch, with the red berries of the mountain-ash shining brightly between them. Through the air above us, like gnomons of a vast sundial, the Aiguilles cast their fanlike shadows, which moved round as the day advanced. Slopes of rhododendrons with withered flowers next succeeded, but the colouring of the bilberry-leaves was scarcely less exquisite than the freshest bloom of the Alpine rose. For a long time we were in the cool shadow of the mountain, catching, at intervals, through the twigs in front of us, glimpses of the sun surrounded by coloured spectra. On one occasion a brow rose in front of me; behind it was a lustrous space of heaven, adjacent to the sun, which, however, was hidden behind the brow; against this space the twigs and weeds upon the summit of the brow shone as if they were self-luminous, while some bits of thistle-down floating in the air appeared, where they crossed this portion of the heavens, like fragments of the sun himself. Once the orb appeared behind a rounded mass of snow which lay near the summit of the Aiguille du Midi. Looked at with the naked eyes, it seemed to possess a billowy motion, the light darting from it in dazzling curves,—a subjective effect produced by the abnormal action of the intense light upon the eye. As the sun's disk came more into view, its rays however still grazing the summit of the mountain, interference-spectra darted from it on all sides, and surrounded it with a glory of richly-coloured bars. Mingling however with the grandeur of nature, we had the anger and obstinacy of man. With a view to subsequent legal proceedings, the Guide Chef sent a spy after us, who, having satisfied himself of our delinquency, took his unpleasant presence from the splendid scene.

Strange to say, though the luminous appearance of bodies projected against the sky adjacent to the rising sun is a most striking and beautiful phenomenon, it is hardly ever seen by either guides or travellers; probably because they avoid looking towards a sky the brightness of which is painful to the eyes. In 1859 Auguste Balmat had never seen the effect; and the only written description of it which we possess is one furnished by Professor Necker, in a letter to Sir David Brewster, which is so interesting that I do not hesitate to reproduce it here:—

"I now come to the point," writes M. Necker, "which you particularly wished me to describe to you; I mean the luminous appearance of trees, shrubs, and birds, when seen from the foot of a mountain a little before sunrise. The wish I had to see again the phenomenon before attempting to describe it made me detain this letter a few days, till I had a fine day to go to see it at the Mont Salève; so yesterday I went there, and studied the fact, and in elucidation of it I made a little drawing, of which I give you here a copy: it will, with the explanation and the annexed diagram (Fig. 9), impart to you, I hope, a correct idea of the phenomenon. You must conceive the observer placed at the foot of a hill interposed between him and the place where the sun is rising, and thus entirely in the shade; the upper margin of the mountain is covered with woods or detached trees and shrubs, which are projected as dark objects on a very bright and clear sky, except at the very place where the sun is just going to rise, for there all the trees and shrubs bordering the margin are entirely,—branches, leaves, stem and all,—of a pure and brilliant white, appearing extremely bright and luminous, although projected on a most brilliant and luminous sky, as that part of it which surrounds the sun always is. All the minutest details, leaves, twigs, &c., are most delicately preserved, and you would fancy you saw these trees and forests made of the purest silver, with all the skill of the most expert workman. The swallows and other birds flying in those particular spots appear like sparks of the most brilliant white. Unfortunately, all these details, which add so much to the beauty of this splendid phenomenon, cannot be represented in such small sketches.

"Neither the hour of the day nor the angle which the object makes with the observer appears to have any effect; for on some occasions I have seen the phenomenon take place at a very early hour in the morning. Yesterday it was 10 A.M., when I saw it as represented in Fig. 10. I saw it again on the same day at 5 P.M., at a different place of the same mountain, for which the sun was just setting. At one time the angle of elevation of the lighted white shrubs above the horizon of the spectator was about 20°, while at another place it was only 15°. But the extent of the field of illumination is variable, according to the distance at which the spectator is placed from it.

When the object behind which the sun is just going to rise, or has just been setting, is very near, no such effect takes place. In the case

Fig. 9.

represented in Fig. 9 the distance was about 194 mètres, or 636 English feet, from the spectator in a direct line, the height above his level being 60 mètres, or 197 English feet, and the horizontal line drawn from him to the horizontal projection of these points on the plane of his horizon being 160 mètres, or 525 English feet, as will be seen in the following diagram, Fig. 10.

" In this case only small shrubs and the lower half of the stem of a tree are illuminated white, and the horizontal

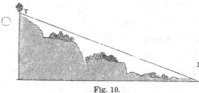

Fig. 10.

extent of this effect is also comparatively small; while at other places when I was near the edge behind which the sun was going to rise no such effect took place. But on the contrary, when I have witnessed the phenomenon at a greater distance and at a greater height, as I have seen it other times on the same and on other mountains of the Alps, large tracts of forests and immense spruce-firs were illuminated white throughout their whole length, as I have attempted to represent in Fig. 11, and the corresponding diagram, Fig. 12. Nothing can be finer than these silver-

looking spruce-forests.   At the same time, though at a
distance of more than a thousand mètres, a vast number of
large swallows or swifts
(*Cypselus alpinus*), which
inhabit these high rocks,
were seen as small bril-
liant stars or sparks moving rapidly in the air.
From these facts it ap-
pears to me obvious that
the extent of the illumi-
nated spots varies in a direct ratio of their distance;
but at the same time that
there must be a constant
angular space, corresponding probably to the zone,
a few minutes of a de-
gree wide, around the
sun's disk, which is a limit

Fig. 11.

to the occurrence of the appearance.   This would explain
how the real extent which it occupies on the earth's surface
varies with the relative distance of the spot from the eye of
the observer, and accounts also for the phenomenon being

Fig. 12.

never seen in the low country, where I have often looked for
it in vain.   Now that you are acquainted with the circumstances of the fact, I have no doubt you will easily observe
it in some part or other of your Scotch hills; it may be

some long heather or furze will play the part of our Alpine forests, and I would advise you to try and place a bee-hive in the required position, and it would perfectly represent our swallows, sparks, and stars."

Our porters, with one exception, reached the Pierre l'Echelle as soon as ourselves; and here having refreshed themselves, and the due exchange of loads having been made, we advanced upon the glacier, which we crossed, until we came nearly opposite to the base of the Grands Mulets. The existence of one wide crevasse, which was deemed impassable, had this year introduced the practice of assailing the rocks at their base, and climbing them to the cabin, an operation which Balmat wished to avoid. At Chamouni, therefore, he had made inquiries regarding the width of the chasm, and acting on his advice I had had a ladder constructed in two pieces, which, united together by iron attachments, was supposed to be of sufficient length to span the fissure. On reaching the latter, the pieces were united, and the ladder thrown across, but the bridge was so frail and shaky at the place of junction, and the chasm so deep, that Balmat pronounced the passage impracticable.

The porters were all grouped beside the crevasse when this announcement was made, and, like hounds in search of the scent, the group instantly broke up, seeking in all directions for a means of passage. The talk was incessant and animating; attention was now called in one direction, anon in another, the men meanwhile throwing themselves into the most picturesque groups and attitudes. All eyes at length were directed upon a fissure which was spanned at one point by an arch of snow, certainly under two feet deep at the crown. A stout rope was tied round the waist of one of our porters, and he was sent forward to test the bridge. He approached it cautiously, treading down the snow to give it compactness, and thus make his footing

sure as he advanced; bringing regelation into play, he
gave the mass the necessary continuity, and crossed in
safety.   The rope was subsequently stretched over the
pont, and each of us causing his right hand to slide along it,
followed without accident.   Soon afterwards, however, we
met with a second and very formidable crevasse, to cross
which we had but half of our ladder, which was applied as
follows:—The side of the fissure on which we stood was
lower than the opposite one; over the edge of the latter
projected a cornice of snow, and a ledge of the same
material jutted from the wall of the crevasse a little below
us.   The ladder was placed from ledge to cornice, both of
its ends being supported by snow.   I could hardly believe
that so frail a bearing could possibly support a man's
weight; but a porter was tied as before, and sent up the
ladder, while we followed protected by the rope.   We
were afterwards tied together, and thus advanced in an
orderly line to the Grands Mulets.

The cabin was wet and disagreeable, but the sunbeams
fell upon the brown rocks outside, and thither Mr. Wills
and myself repaired to watch the changes of the atmosphere.   I took possession of the flat summit of a prism of
rock, where, lying upon my back, I watched the clouds
forming, and melting, and massing themselves together,
and tearing themselves like wool asunder in the air above.
It was nature's language addressed to the intellect;
these clouds were visible symbols which enabled us to
understand what was going on in the invisible air.   Here
unseen currents met, possessing different temperatures,
mixing their contents both of humidity and motion, producing a mean temperature unable to hold their moisture
in a state of vapour.   The water-particles, obeying their
mutual attractions, closed up, and a visible cloud suddenly
shook itself out, where a moment before we had the pure
blue of heaven.   Some of the clouds were wafted by the

air towards atmospheric regions already saturated with
moisture, and along their frontal borders new cloudlets
ever piled themselves, while the hinder portions, invaded
by a drier or a warmer air, were dissipated; thus the
cloud advanced, with gain in front and loss behind, its permanence depending on the balance between them. The
day waned, and the sunbeams began to assume the colouring due to their passage through the horizontal air. The
glorious light, ever deepening in colour, was poured boun-
teously over crags, and snows, and clouds, and suffused
with gold and crimson the atmosphere itself. I had never
seen anything grander than the sunset on that day.
Clouds with their central portions densely black, denying
all passage to the beams which smote them, floated
westward, while the fiery fringes which bordered them
were rendered doubly vivid by contrast with the adjacent
gloom. The smaller and more attenuated clouds were
intensely illuminated throughout. Across other inky
masses were drawn zigzag bars of radiance which resembled streaks of lightning. The firmament between
the clouds faded from a blood-red through orange and
daffodil into an exquisite green, which spread like a
sea of glory through which those magnificent argosies
slowly sailed. Some of the clouds were drawn in straight
chords across the arch of heaven, these being doubtless
the sections of layers of cloud whose horizontal dimensions were hidden from us. The cumuli around and near
the sun himself could not be gazed upon, until, as the
day declined, they gradually lost their effulgence and
became tolerable to the eyes. All was calm—but there
was a wildness in the sky like that of anger, which boded
evil passions on the part of the atmosphere. The sun
at length sank behind the hills, but for some time afterwards carmine clouds swung themselves on high, and cast
their ruddy hues upon the mountain snows. Duskier and

colder waxed the west, colder and sharper the breeze
of evening upon the Grands Mulets, and as twilight
deepened towards night, and the stars commenced to
twinkle through the chilled air, we retired from the
scene.

The anticipated storm at length gave notice of its
coming. The sea-waves, as observed by Aristotle, sometimes reach the shore before the wind which produces them
is felt; and here the tempest sent out its precursors,
which broke in detached shocks upon the cabin before the
real storm arrived. Billows of air, in ever quicker succession, rolled over us with a long surging sound, rising and
falling as crest succeeded trough and trough succeeded
crest. And as the pulses of a vibrating body, when
their succession is quick enough, blend to a continuous
note, so these fitful gusts linked themselves finally to a
storm which made its own wild music among the crags.
Grandly it swelled, carrying the imagination out of doors,
to the clouds and darkness, to the loosened avalanches and
whirling snow upon the mountain heads. Moored to the rock
on two sides, the cabin stood firm, and its manifest security
allowed the mind the undisturbed enjoyment of the atmospheric war. We were powerfully shaken, but had no
fear of being uprooted; and a certain grandeur of the
heart rose responsive to the grandeur of the storm.
Mounting higher and higher, it at length reached its
maximum strength, from which it lowered fitfully, until
at length, with a melancholy wail, it bade our rock farewell.

A little before half-past one we issued from the cabin.
The night being without a moon, we carried three lanterns.
The heavens were crowded with stars, among which, how-
ever, angry masses of cloud here and there still wandered.
The storm, too, had left a rear-guard behind it; and strong
gusts rolled down upon us at intervals, at one time, indeed,

so violent as to cause Balmat to express doubts of our
being able to reach the summit. With a thick handkerchief
bound around my hat and ears I enjoyed the onset of the
wind. Once, turning my head to the left, I saw what appeared to me to be a huge mass of stratus cloud, at a
great distance, with the stars shining over it. In another
instant a precipice of *névé* loomed upon us ; we were close
to its base, and along its front the annual layers were separated from each other by broad dark bands. Through the
gloom it appeared like a cloud, the lines of bedding giving
to it the stratus character.

Immediately before lying down on the previous evening
I had opened the little window of the cabin to admit some
air. In the sky in front of me shone a curious nodule of misty
light with a pale train attached to it. In 1853, on the
side of the Brocken, I had observed, without previous
notice, a comet discovered a few days previously by a
former fellow student, and here was another "discovery"
of the same kind. I inspected the stranger with my
telescope, and assured myself that it was a comet. Mr.
Wills chanced to be outside at the time, and made the same
observation independently. As we now advanced up the
mountain its ominous light gleamed behind us, while high
up in heaven to our left the planet Jupiter burned like
a lamp of intense brightness. The Petit Plateau forms
a kind of reservoir for the avalanches of the Dôme du
Goûté, and this year the accumulation of frozen débris
upon it was enormous. We could see nothing but the ice-
blocks on which the light of the lanterns immediately fell ;
we only knew that they had been discharged from the
séracs, and that similar masses now rose threatening to our
right, and might at any moment leap down upon us. Balmat commanded silence, and urged us to move across the
plateau with all possible celerity. The warning of our
guide, the wild and rakish appearance of the sky, the spent

projectiles at our feet, and the comet with its "horrid hair" behind, formed a combination eminently calculated to excite the imagination.

And now the sky began to brighten towards dawn, with that deep and calm beauty which suggests the thought of adoration to the human mind. Helped by the contemplation of the brightening east, which seemed to lend lightness to our muscles, we cheerily breasted the steep slope up to the Grand Plateau. The snow here was deep, and each of our porters took the lead in turn. We paused upon the Grand Plateau and had breakfast; digging, while we halted, our feet deeply into the snow. Thence up to the corridor, by a totally different route from that pursued by Mr. Hirst and myself the year previously; the slope was steep, but it had not a precipice for its boundary. Deep steps were necessary for a time, but when we reached the summit our ascent became more gentle. The eastern sky continued to brighten, and by its illumination the Grand Plateau and its bounding heights were lovely beyond conception. The snow was of the purest white, and the glacier, as it pushed itself on all sides into the basin, was riven by fissures filled with a cœrulean light, which deepened to inky gloom as the vision descended into them. The edges were overhung with fretted cornices, from which depended long clear icicles, tapering from their abutments like spears of crystal. The distant fissures, across which the vision ranged obliquely without descending into them, emitted that magical firmamental shimmer which, contrasted with the pure white of the snow, was inexpressibly lovely. Near to us also grand castles of ice reared themselves, some erect, some overturned, with clear cut sides, striped by the courses of the annual snows, while high above the *séracs* of the plateau rose their still grander brothers of the Dôme du Goûté. There was a nobility in this glacier scene which I think I have never seen surpassed;—a

strength of nature, and yet a tenderness, which at once
raised and purified the soul. The gush of the direct
sunlight could add nothing to this heavenly beauty; indeed I thought its yellow beams a profanation as they
crept down from the humps of the Dromedary, and invaded
more and more the solemn purity of the realm below.

Our way lay for a time amid fine fissures with blue
walls, until at length we reached the edge of one which
elicited other sentiments than those of admiration. It
must be crossed. At the opposite side was a high and
steep bank of ice which prolonged itself downwards, and
ended in a dependent eave of snow which quite overhung the chasm, and reached to within about a yard of
our edge of the crevasse. Balmat came forward with his
axe, and tried to get a footing on the eave : he beat it
gently, but the axe went through the snow, forming an
aperture through which the darkness of the chasm was
rendered visible. Our guide was quite free, without rope
or any other means of security ; he beat down the snow so
as to form a kind of stirrup, and upon this he stepped.
The stirrup gave way, it was right over the centre of the
chasm, but with wonderful tact and coolness he contrived
to get sufficient purchase from the yielding mass to toss
himself back to the side of the chasm. The rope was
now brought forward and tied round the waist of one
of the porters; another step was cautiously made in the
eave of snow, the man was helped across, and lessened
his own weight by means of his hatchet. He gradually got footing on the face of the steep, which he
mounted by escaliers; and on reaching a sufficient height
he cut two large steps in which his feet might rest
securely. Here he laid his breast against the sloping wall,
and another person was sent forward, who drew himself up
by the rope which was attached to the leader. Thus we all
passed, each of us in turn bearing the strain of his succes-

sor upon the rope; it was our last difficulty, and we afterwards slowly plodded through the snow of the corridor
towards the base of the Mur de la Côte.

Climbing zigzag, we soon reached the summit of
the Mur, and immediately afterwards found ourselves
in the midst of cold drifting clouds, which obscured
everything. They dissolved for a moment and revealed to us the sunny valley of Chamouni; but they
soon swept down again and completely enveloped us.
Upon the Calotte, or last slope, I felt no trace of the exhaustion which I had experienced last year, but enjoyed free
lungs and a quiet heart. The clouds now whirled wildly
round us, and the fine snow, which was caught by
the wind and spit bitterly against us, cut off all visible
communication between us and the lower world. As
we approached the summit the air thickened more and
more, and the cold, resulting from the withdrawal of the
sunbeams, became intense. We reached the top, however, in good condition, and found the new snow piled up
into a sharp *arête*, and the summit of a form quite different from that of the *Dos d'un Ane*, which it had presented
the previous year. Leaving Balmat to make a hole for the
thermometer, I collected a number of batons, drove them
into the snow, and, drawing my plaid round them, formed
a kind of extempore tent to shelter my boiling-water
apparatus. The covering was tightly held, but the snow
was as fine and dry as dust, and penetrated everywhere:
my lamp could not be secured from it, and half a box of
matches was consumed in the effort to ignite it. At length
it did flame up, and carried on a sputtering combustion.
The cold of the snow-filled boiler condensing the vapour
from the lamp gradually produced a drop, which, when
heavy enough to detach itself from the vessel, fell upon
the flame and put it out. It required much patience and
the expenditure of many matches to relight it. Meanwhile

the absence of muscular action caused the cold to affect our men severely. My beard and whiskers were a mass of clotted ice. The batons were coated with ice, and even the stem of my thermometer, the bulb of which was in hot water, was covered by a frozen enamel. The clouds whirled, and the little snow granules hit spitefully against the skin wherever it was exposed. The temperature of the air was 20° Fahr. below the freezing point. I was too intent upon my work to heed the cold much, but I was numbed; one of my fingers had lost sensation, and my right heel was in pain: still I had no thought of forsaking my observation until Mr. Wills came to me and said that we must return speedily, for Balmat's hands were *gelées*. I did not comprehend the full significance of the word; but, looking at the porters, they presented such an aspect of suffering that I feared to detain them longer. They looked like worn old men, their hair and clothing white with snow, and their faces blue, withered, and anxious-looking. The hole being ready, I asked Balmat for the magnet to arrange the index of the thermometer: his hands seemed powerless. I struck my tent, deposited the instrument, and, as I watched the covering of it up, some of the party, among whom were Mr. Wills and Balmat, commenced the descent.*

I followed them speedily. Midway down the Calotte I saw Balmat, who was about a hundred yards in advance of me, suddenly pause and thrust his hands into the snow, and commence rubbing them vigorously. The suddenness of the act surprised me, but I had no idea at the time of its real significance: I soon came up to him; he seemed frightened, and continued to beat and rub his hands, plunging them, at quick intervals, into the snow. Still

---

* In August, 1859, I found the temperature of water, boiling in an open vessel at the summit of Mont Blanc, to be 184·95° Fahr. On that occasion also, though a laborious search was made for the thermometer, it could not be found.

I thought the thing would speedily pass away, for I had too much faith in the man's experience to suppose that he would permit himself to be seriously injured. But it did not pass as I hoped it would, and the terrible possibility of his losing his hands presented itself to me. He at length became exhausted by his own efforts, staggered like a drunken man, and fell upon the snow. Mr. Wills and myself took each a hand, and continued the process of beating and rubbing. I feared that we should injure him by our blows, but he continued to exclaim, "N'ayez pas peur, frappez toujours, frappez fortement!" We did so, until Mr. Wills became exhausted, and a porter had to take his place. Meanwhile Balmat pinched and bit his fingers at intervals, to test their condition; but there was no sensation. He was evidently hopeless himself; and, seeing him thus, produced an effect upon me that I had not experienced since my boyhood—my heart swelled, and I could have wept like a child. The idea that I should be in some measure the cause of his losing his hands was horrible to me; schemes for his support rushed through my mind with the usual swiftness of such speculations, but no scheme could restore to him his lost hands. At length returning sensation in one hand announced itself by excruciating pain. "Je souffre!" he exclaimed at intervals—words which, from a man of his iron endurance, had a more than ordinary significance. But pain was better than death, and, under the circumstances, a sign of improvement. We resumed our descent, while he continued to rub his hands with snow and brandy, thrusting them at every few paces into the mass through which we marched. At Chamouni he had skilful medical advice, by adhering to which he escaped with the loss of six of his nails—his hands were saved.

I cannot close this recital without expressing my admiration of the dauntless bearing of our porters, and of the cheerful and efficient manner in which they did their duty throughout the whole expedition. Their names are Edouard Bellin, Joseph Favret, Michel Payot, Joseph Folliguet, and Alexandre Balmat.

==========

(26.)

The hostility of the chief guide to the expedition was not diminished by the letter of the Intendant; and he at once entered a *procès verbal* against Balmat and his companions on their return to Chamouni. I felt that the power thus vested in an unlettered man to arrest the progress of scientific observations was so anomalous, that the enlightened and liberal Government of Sardinia would never tolerate such a state of things if properly represented to it. The British Association met at Leeds that year, and to it, as a guardian of science, my thoughts turned. I accordingly laid the case before the Association, and obtained its support: a resolution was unanimously passed " that application be made to the Sardinian authorities for increased facilities for making scientific observations in the Alps."

Considering the arduous work which Balmat had performed in former years in connexion with the glaciers, and especially his zeal in determining, under the direction of Professor Forbes, their winter motion—for which, as in the case above recorded, he refused all personal remuneration —I thought such services worthy of some recognition on the part of the Royal Society. I suggested this to the Council, and was met by the same cordial spirit of co-operation which I had previously experienced at Leeds. A sum of five-and-twenty guineas was at once voted for the purchase of a suitable testimonial; and a committee, consisting of Sir Roderick Murchison, Professor Forbes, and myself, was appointed to carry the thing out. Balmat was consulted, and he chose a photographic apparatus, which, with a suitable inscription, was duly presented to him.

Thus fortified, I drew up an account of what had occurred at Chamouni during my last visit, accompanied by a brief statement of the changes which seemed desirable. This was placed in the hands of the President of the British Association, to whose prompt and powerful cooperation in this matter every Alpine explorer who aspires to higher ground than ordinary is deeply indebted. The following letter assured me that the facility applied for by the British Association would be granted by the Sardinian Government, and that future men of science would find in the Alps a less embarrassed field of operations than had fallen to my lot in the summer of 1858.

"12, Hertford-street, Mayfair, W.,
"My dear Sir,—   "February 18th, 1859.

"Having, as I informed you in my last note, communicated with the Sardinian Minister Plenipotentiary the day after receiving your statement relative to the guides at Chamouni, I have been favoured by replies from the Minister, of the 4th and 17th February. In the first the Marquis d'Azeglio assures me that he will bring the subject before the competent authorities at Turin, accompanying the transmission 'd'une récommendation toute spéciale.' In the second letter the Marquis informs me that 'the preparation of new regulations for the guides at Chamouni had for some time occupied the attention of the Minister of the Interior, and that thesé regulations will be in rigorous operation, in all probability, at the commencement of the approaching summer.' The Marquis adds that, 'as the regulations will be based upon a principle of much greater liberty, he has every reason to believe that

K

they will satisfy all the desires of travellers in the interests of science.'

"With much pleasure at the opportunity of having been in any degree able to bring about the fulfilment of your wishes on the subject,

"I remain, my dear Sir,

"Faithfully yours,

"RICHARD OWEN.

"Pres. Brit. Association.

"Prof. Tyndall, F.R.S."

It ought to be stated, that, previous to my arrival at Chamouni in 1858, an extremely cogent memorial drawn up by Mr. John Ball had been presented to the Marquis d'Azeglio by a deputation from the Alpine Club. It was probably this memorial which first directed the attention of the Sardinian Minister of the Interior to the subject.

WINTER EXPEDITION TO THE MER DE GLACE, 1859.

(27.)

HAVING ten days at my disposal last Christmas, I was anxious to employ them in making myself acquainted with the winter aspects and phenomena of the Mer de Glace. On Wednesday, the 21st of December, I accordingly took my place to Paris, but on arriving at Folkestone found the sea so tempestuous that no boat would venture out.

The loss of a single day was more than I could afford, and this failure really involved the loss of two. Seeing, therefore, the prospect of any practical success so small, I returned to London, purposing to give the expedition up. On the following day, however, the weather lightened, and I started again, reaching Paris on Friday morning. On that day it was not possible to proceed beyond Macon, where, accordingly, I spent the night, and on the following day reached Geneva.

Much snow had fallen; at Paris it still cumbered the streets, and round about Macon it lay thick, as if a more than usually heavy cloud had discharged itself on that portion of the country. Between Macon and Roussillon it was lighter, but from the latter station onwards the quantity upon the ground gradually increased.

On Christmas morning, at 8 o'clock, I left Geneva by the diligence for Sallenches. The dawn was dull, but the sky cleared as the day advanced, and finally a dome of cloudless blue stretched overhead. The mountains were grand; their sunward portions of dazzling whiteness, while the shaded sides, in contrast with the blue sky behind them, presented a ruddy, subjective tint. The brightness of the

K 2

day reached its maximum towards one o'clock, after which a milkiness slowly stole over the heavens, and increased in density until finally a drowsy turbidity filled the entire air. The distant peaks gradually blended with the white atmosphere above them and lost their definition. The black pine forests on the slopes of the mountains stood out in strong contrast to the snow; and, when looked at through the spaces enclosed by the tree branches at either side of the road, they appeared of a decided indigo-blue. It was only when thus detached by a vista in front that the blue colour was well seen, the air itself between the eye and the distant pines being the seat of the colour. Goethe would have regarded it as an excellent illustration of his 'Farbenlehre.'

We reached Sallenches a little after 4 p. m., where I endeavoured to obtain a sledge to continue my journey. A fit one was not to be found, and a carriage was therefore the only resort. We started at five; it was very dark, but the feeble reflex of the snow on each side of the road was preferred by the postilion to the light of lamps. Unlike the enviable ostrich, I cannot shut my eyes to danger when it is near: and as the carriage swayed towards the precipitous road side, I could not fold myself up, as it was intended I should, but, quitting the interior and divesting my limbs of every encumbrance, I took my seat beside the driver, and kept myself in readiness for the spring, which in some cases appeared imminent. My companion however was young, strong, and keen-eyed; and though we often had occasion for the exercise of the quality last mentioned, we reached Servoz without accident.

Here we baited, and our progress afterwards was slow and difficult. The snow on the road was deep and hummocky, and the strain upon the horses very great. Having crossed the Arve at the Pont-Pelissier, we both alighted, and I went on in advance. The air was

warm, and not a whisper disturbed its perfect repose.
There was no moon, and the heavy clouds, which now
quite overspread the heavens, cut off even the feeble
light of the stars. The sound of the Arve, as it rushed
through the deep valley to my left, came up to me
through crags and trees with a sad murmur. Sometimes on passing an obstacle, the sound was entirely cut
off, and the consequent silence was solemn in the extreme.
It was a churchyard stillness, and the tall black pines,
which at intervals cast their superadded gloom upon the
road, seemed like the hearse-plumes of a dead world. I
reached a wooden hut, where a lame man offers batons,
minerals, and *eau de vie*, to travellers in summer. It was
forsaken, and half buried in the snow. I leaned against
the door, and enjoyed for a time the sternness of the
surrounding scene. My conveyance was far behind, and
the intermittent tinkle of the horses' bells, which augmented instead of diminishing the sense of solitude, informed me of the progress and the pauses of the vehicle.
At the summit of the road I halted until my companion
reached me ; we then both remounted, and proceeded slowly
towards Les Ouches. We passed some houses, the aspect
of which was even more dismal than that of Nature ; their
roofs were loaded with snow, and white buttresses were
reared against the walls. There was no sound, no light,
no voice of joy to indicate that it was the pleasant Christmas
time. We once met the pioneer of a party of four drunken
peasants : he came right against us, and the coachman had
to pull up. Planting his feet in the snow and propping
himself against the leader's shoulder, the bacchanal
exhorted the postilion to drive on; the latter took him
at his word, and overturned him in the snow. After
this we encountered no living thing. The horses seemed
seized by a kind of torpor, and leaned listlessly against
each other ; vainly the postilion endeavoured to rouse

them by word and whip; they sometimes essayed to
trot down the slopes, but immediately subsided to their
former monotonous crawl. As we ascended the valley,
the stillness of the air was broken at intervals by wild
storm-gusts, sent down against us from Mont Blanc
himself. These chilled me, so I quitted the carriage,
and walked on. Not far from Chamouni, the road, for
some distance, had been exposed to the full action of the
wind, and the snow had practically erased it. Its left wall
was completely covered, while a few detached stones,
rising here and there above the surface, were the only
indications of the presence and direction of the right hand
wall. I could not see the state of the surface, but I
learned by other means that the snow had been heaped
in oblique ridges across my path. I staggered over four
or five of these in succession, sinking knee-deep, and
finally found myself immersed to the waist. This made
me pause; I thought I must have lost the road, and vainly
endeavoured to check myself by the positions of surrounding objects. I turned back and met the carriage: it had
stuck in one of the ridges; one horse was down, his hind
legs buried to the haunches, his left fore leg plunged to
the shoulder in snow, and the right one thrown forward
upon the surface. *C'est bien la route?* demanded my companion. I went back exploring, and assured myself that
we were over the road; but I recommended him to release
the horses and leave the carriage to its fate. He, however,
succeeded in extricating the leader, and while I went on
in advance seeking out the firmer portions of the road, he
followed, holding his horses by their heads; and half an
hour's struggle of this kind brought us to Chamouni.

It also was a little "city of the dead." There was no
living thing in the streets, and neither sound nor light in
the houses. The fountain made a melancholy gurgle, one or
two loosened window-shutters creaked harshly in the wind,

and banged against the objects which limited their oscillations. The hotel de l'Union, so bright and gay in summer, was nailed up and forsaken; and the cross in front of it, stretching its snow-laden arms into the dim air, was the type of desolation. We rang the bell at the Hôtel Royal, but the bay of a watch-dog resounding through the house was long our only reply. The bell appeared powerless to wake the sleepers, and its sound mingled dismally with that of the wind howling through the deserted passages. The noise of my boot-heel, exerted long on the front door, was at length effective; it was unbarred, and the physical heat of a good stove soon added itself to the warmth of the welcome with which my hostess greeted me.

December 26th.—The snow fell heavily, at frequent intervals, throughout the entire day. Dense clouds draped all the mountains, and there was not the least prospect of my being able to see across the Mer de Glace. I walked out alone in the dim light, and afterwards traversed the streets before going to bed. They were quite forsaken. Cold and sullen the Arve rolled under its wooden bridge, while the snow fell at intervals with heavy shock from the roofs of the houses, the partial echos from the surfaces of the granules combining to render the sound loud and hollow. Thus the concerns of this little hamlet were changed and fashioned by the obliquity of the earth's axis, the chain of dependence which runs throughout creation, linking the roll of a planet alike with the interests of marmots and of men.

Tuesday, 27th December.—I rose at six o'clock, having arranged with my men to start at seven, if the weather at all permitted. Edouard Simond, my old assistant of 1857, and Joseph Tairraz were the guides of the party; the porters were Edouard Balmat, Joseph Simond (fils d'Auguste), Francois Ravanal, and another. They came at the time appointed; it was snowing heavily, and we agreed to wait till eight o'clock and then decide. They returned at eight, and finding them disposed to try the ascent to the Montanvert, it was not my place to baulk them. Through the valley the work was easy, as the snow had been partially beaten down, but we soon passed the habitable limits, and had to break ground for ourselves. Three of my men had tried to reach the Montanvert by *la Filia* on the previous Thursday, but their experience of the route had been such as to deter them from trying it again. We now chose the ordinary route, breasting the slope until we reached the cluster of chalets, under the projecting eave of one of which the men halted and applied "pattens" to their feet. These consisted of planks about sixteen inches long and ten wide, which were firmly strapped to the feet. My first impression was that they were worse than useless, for though they sank less deeply than the unarmed feet, on being raised they carried with them a larger amount of snow, which, with the leverage of the leg, appeared to necessitate an enormous waste of force. I stated this emphatically, but the men adhered to their pattens, and before I reached the Montanvert I had reason to commend their practice as preferable to my theory. I was however guided by the latter, and wore no pattens. The general depth of the snow along the track was over three feet; the footmarks of the men were usually rigid enough to bear my weight, but in many cases I went through the crust which their pressure had produced, and sank suddenly in the mass. The snow became softer as we ascended, and my immersions more frequent, but the work was pure enjoyment, and the scene one of extreme beauty. The previous night's snow had descended through a perfectly still atmosphere, and had loaded all the branches of the pines; the long arms of the trees drooped under the weight, and presented at their extremities the appearance of enormous talons turned downwards. Some

of the smaller and thicker trees were almost entirely
covered, and assumed grotesque and beautiful forms; the
upper part of one in particular resembled a huge white
parrot with folded wings and drooping head, the slumber
of the bird harmonizing with the torpor of surrounding
nature. I have given a sketch of it in Fig. 13.

Fig. 13.

Previous to reaching the half-way spring, where the
peasant girls offer strawberries to travellers in summer,
we crossed two large couloirs filled with the débris of
avalanches which had fallen the night before. Between
these was a ridge forty or fifty yards wide on which the
snow was very deep, the slope of the mountain also adding
a component to the fair thickness of the snow. My
shoulder grazed the top of the embankment to my right as
I crossed the ridge, and once or twice I found myself waist
deep in a vertical shaft from which it required a consider-

able effort to escape. Suddenly we heard a deep sound resembling the dull report of a distant gun, and at the same moment the snow above us broke across, forming a fissure parallel to our line of march. The layer of snow had been in a state of strain, which our crossing brought to a crisis: it gave way, but having thus relieved itself it did not descend. Several times during the ascent the same phenomenon occurred. Once, while engaged upon a very steep slope, one of the men cried out to the leader, *"Arrêtez!"* Immediately in front of the latter the snow had given way, forming a zigzag fissure across the slope. We all paused, expecting to see an avalanche descend. Tairraz was in front; he struck the snow with his baton to loosen it, but seeing it indisposed to descend he advanced cautiously across it, and was followed by the others. I brought up the rear. The steepness of the mountain side at this place, and the absence of any object to which one might cling, would have rendered a descent with the snow in the last degree perilous, and we all felt more at ease when a safe footing was secured at the further side of the incline.

At the spring, which showed a little water, the men paused to have a morsel of bread. The wind had changed, the air was clearing, and our hopes brightening. As we ascended the atmosphere went through some extraordinary mutations. Clouds at first gathered round the Aiguille and Dôme du Gouté, casting the lower slopes of the mountain into intense gloom. After a little time all this cleared away, and the beams of the sun striking detached pieces of the slopes and summits produced an extraordinary effect. The Aiguille and Dôme were most singularly illumined, and to the extreme left rose the white conical hump of the Dromedary, from which a long streamer of snow-dust was carried southward by the wind. The Aiguille du Dru, which had been completely mantled during the earlier part of the day, now threw off its cloak

of vapour and rose in most solemn majesty before us;
half of its granite cone was warmly illuminated, and
half in shadow. The wind was high in the upper regions,
and, catching the dry snow which rested on the asperities
and ledges of the Aiguille, shook it out like a vast banner
in the air. The changes of the atmosphere, and the
grandeur which they by turns revealed and concealed,
deprived the ascent of all weariness. We were usually
flanked right and left by pines, but once between the
fountain and the Montanvert we had to cross a wide unsheltered portion of the mountain which was quite covered
with the snow of recent avalanches. This was lumpy and
far more coherent than the undisturbed snow. We took
advantage of this, and climbed zigzag over the avalanches
for three-quarters of an hour, thus reaching the opposite
pines at a point considerably higher than the path. This,
though not the least dangerous, was the least fatiguing
part of the ascent.

I frequently examined the colour of the snow: though
fresh, its blue tint was by no means so pronounced as I
have seen it on other occasions; still it was beautiful. The
colour is, no doubt, due to the optical reverberations
which occur within a fissure or cavity formed in the
snow. The light is sent from side to side, each time
plunging a little way into the mass; and being ejected
from it by reflection, it thus undergoes a sifting process,
and finally reaches the eye as blue light. The presence of any object which cuts off this cross-fire of the
light destroys the colour. I made conical apertures in the
snow, in some cases three feet deep, a foot wide at the
mouth, and tapering down to the width of my baton.
When the latter was placed along the axis of such a cone,
the blue light which had previously filled the cavity disappeared; on the withdrawal of the baton it was followed
by the light, and thus by moving the staff up and down

its motions were followed by the alternate appearance and extinction of the light. I have said that the holes made in the snow seemed filled with a blue light, and it certainly appeared as if the air contained in the cavities had itself been coloured, and thereby rendered visible, the vision plunging into it as into a blue medium. Another fact is perhaps worth notice : snow rarely lies so smooth as not to present little asperities at its surface ; little ridges or hillocks, with little hollows between them. Such small hollows resemble, in some degree, the cavities which I made in the snow, and from them, in the present instance, a delicate light was sent to the eye, faintly tinted with the pure blue of the snow-crystals. In comparison with the spots thus illuminated, the little protuberances were gray. The portions most exposed to the light seemed least illuminated, and their defect in this respect made them appear as if a light-brown dust had been strewn over them.

After five hours and a half of hard work we reached the Montanvert. I had often seen it with pleasure. Often, having spent the day alone amid the *séracs* of the Col du Géant, on turning the promontory of Trelaporte on my way home, the sight of the little mansion has gladdened me, and given me vigour to scamper down the glacier, knowing that pleasant faces and wholesome fare were awaiting me. This day, also, the sight of it was most welcome, despite its desolation. The wind had swept round the auberge, and carried away its snow-buttresses, piling the mass thus displaced against the adjacent sheds, to the roofs of which one might step from the surface of the snow. The floor of the little château in which I lodged in 1857 was covered with snow, and on it were the fresh footmarks of a little animal—a marmot might have made such marks, had not the marmots been all asleep—what the creature was I do not know.

In the application of her own principles, Nature often transcends the human imagination; her acts are bolder than our predictions. It is thus with the motion of glaciers; it was thus at the Montanvert on the day now referred to. The floors, even where the windows appeared well closed, were covered with a thin layer of fine snow; and some of the mattresses in the bedrooms were coated to the depth of half an inch with this fine powder. Given a chink through which the finest dust can pass, dry snow appears competent to make its way through the same fissure. It had also been beaten against the windows, and clung there like a ribbed drapery. In one case an effect so singular was exhibited, that I doubted my eyes when I first saw it. In front of a large pane of glass, and quite detached from it, save at its upper edge, was a festooned curtain formed entirely of minute ice crystals. It appeared to be as fine as muslin; the ease of its curves and the depth of its folds being such as could not be excelled by the intentional arrangement of ordinary gauze. The frost-figures on some of the window-panes were also of the most extraordinary character: in some cases they extended over large spaces, and presented the appearance which we often observe in London; but on other panes they occurred in detached clusters, or in single flowers, these grouping themselves together to form miniature bouquets of inimitable beauty. I placed my warm hand against a pane which was covered by the crystallization, and melted the frostwork which clung to it. I then withdrew my hand and looked at the film of liquid through a pocket-lens. The glass cooled by contact with the air, and after a time the film commenced to move at one of its edges; atom closed with atom, and the motion ran in living lines through the pellicle, until finally the entire film presented the beauty and delicacy of an organism. The connexion between such objects and what we are accustomed to call the feelings may not be manifest, but it is nevertheless true that, besides appealing to the pure intellect of man, these exquisite productions can also gladden his heart and moisten his eyes.

The glacier excited the admiration of us all: not as in summer, shrunk and sullied like a spent reptile, steaming under the influence of the sun, its frozen muscles were compact; strength and beauty were associated in its aspect. At some places it was pure and smooth; at others frozen fins arose from it, high, steep, and sharply crested. Down the opposite mountain side arrested streams set themselves erect in successive terraces, the fronts of which were fluted pillars of ice. There was no sound of water; even the Nant Blanc, which gushes from a spring, and which some describe as permanent throughout the winter, showed no trace of existence. From the Montanvert to Trelaporte the Mer de Glace was all in shadow; but the sunbeams pouring down the corridor of the Géant ruled a beam of light across the glacier at its upper portion, smote the base of the Aiguille du Moine, and flooded the mountain with glory to its crest. At the opposite side of the valley was the Aiguille du Dru, with a banneret of snow streaming from its mighty cone. The Grande Jorasse, and the range of summits between it and the Aiguille du Géant, were all in view, and the Charmoz raised its precipitous cliffs to the right, and pierced with its splinter-like pinnacles the clear cold air. As the night drew on, the mountains seemed to close in upon us; and on looking out before retiring to rest, a scene so solemn had never before presented itself to my eyes or affected my imagination.

My men occupied the afternoon of the day of our arrival in making a preliminary essay upon the glacier while I prepared my instruments. To the person whom I intended to fix my stations, three others were at-

tached by sound ropes of considerable length. Hidden
crevasses we knew were to be encountered, and we had
made due preparation for them. Throughout the afternoon
the weather remained fine, and at night the stars shone
out, but still with a feeble lustre. I could notice a turbidity gathering in the air over the range of the Brévent,
which seemed disposed to extend itself towards us. At
night I placed a chair in the middle of the snow, at some
distance from the house, and laid on it a registering
thermometer. A bountiful fire of pine logs was made in
the *salle à manger;* a mattress was placed with its foot
towards the fire, its middle line bisecting the right angle
in which the fireplace stood; this being found by experiment to be the position in which the drafts from the door
and from the windows most effectually neutralized each
other. In this region of calms I lay down, and covering
myself with blankets and duvets, listened to the crackling of the logs, and watched their ruddy flicker upon the
walls, until I fell asleep.

The wind rose during the night, and shook the windows:
one pane in particular seemed set in unison to the gusts,
and responded to them by a loud and melodious vibration.
I rose and wedged it round with *sous* and penny pieces,
and thus quenched its untimely music.

December 28th.—We were up before the dawn.
Tairraz put my fire in order, and I then rose. The temperature of the room at a distance of eight feet from
the fire was two degrees of Centigrade below zero; the
lowest temperature outside was eleven degrees of Centigrade below zero,—not at all an excessive cold. The
clouds indeed had, during the night, thrown vast diaphragms
across the sky, and thus prevented the escape of the earth's
heat into space.

While my assistants were preparing breakfast I had
time to inspect the glacier and its bounding heights. On

looking up the Mer de Glace, the Grande Jorasse meets the view, rising in steep outline from the wall of cliffs which terminates the Glacier de Léchaud. Behind this steep ascending ridge, which is shown on the frontispiece, and upon it, a series of clouds had ranged themselves, stretching lightly along the ridge at some places, and at others collecting into ganglia. A string of rosettes was thus formed which were connected together by gauzy filaments. The portion of the heavens behind the ridge was near the domain of the rising sun, and when he cleared the horizon his red light fell upon the clouds, and ignited them to ruddy flames. Some of the lighter clouds doubled round the summit of the mountain, and swathed its black crags with a vestment of transparent red. The adjacent sky wore a strange and supernatural air; indeed there was something in the whole scene which baffled analysis, and the words of Tennyson rose to my lips as I gazed upon it :—

"God made Himself an awful rose of dawn."

I have spoken several times of the cloud-flag which the wind wafted from the summit of the Aiguille du Dru. On the present occasion this grand banner reached extraordinary dimensions. It was brindled in some places as if whipped into curds by the wind; but through these continuous streamers were drawn, which were bent into sinuosities resembling a waving flag at a mast-head. All this was now illuminated with the sun's red rays, which also fringed with fire the exposed edges and pinnacles both of the Aiguille du Dru and the Aiguille Verte. Thus rising out of the shade of the valley the mountains burned like a pair of torches, the flames of which were blown half a mile through the air. Soon afterwards the summits of the Aiguilles Rouges were illuminated, and day declared itself openly among the mountains.

But these red clouds of the morning, magnificent though they were, suggested thoughts which tended to qualify the pleasure which they gave: they did not indicate good weather. Sometimes, indeed, they had to fight with denser masses, which often prevailed, swathing the mountains in deep neutral tint, but which, again yielding, left the glory of the sunrise augmented by contrast with their gloom. Between eight and nine A.M. we commenced the setting out of our first line, one of whose termini was a point about a hundred yards higher up than the Montanvert hotel; a withered pine on the opposite mountain side marking the other terminus. The stakes made use of were four feet long. With the selfsame baton which I had employed upon the Mer de Glace in 1857, and which Simond had preserved, the worthy fellow now took up the line. At some places the snow was very deep, but its lower portions were sufficiently compact to allow of a stake being firmly fixed in it. At those places where the wind had removed the snow or rendered it thin, the ice was pierced with an auger and the stake driven into it. The greatest caution was of course necessary on the part of the men; they were in the midst of concealed crevasses, and sounding was essential at every step. By degrees they withdrew from me, and approached the eastern boundary of the glacier, where the ice was greatly dislocated, and the labour of wading through the snow enormous. Long détours were sometimes necessary to reach a required point; but they were all accomplished, and we at length succeeded in fixing eleven stakes along this line, the most distant of which was within about eighty yards of the opposite side of the glacier.

The men returned, and I consulted them as to the possibility of getting a line across at the *Ponts;* but this was judged to be impossible in the time. We thought, however, that a second line might be staked out at some distance below the Montanvert. I took the theodolite down the mountain-slope, wading at times breast deep in snow, and having selected a line, the men tied themselves together as before, and commenced the staking out. The work was slowly, but steadily and steadfastly done. The air darkened; angry clouds gathered around the mountains, and at times the glacier was swept by wild squalls. The men were sometimes hidden from me by the clouds of snow which enveloped them, but between those intermittent gusts there were intervals of repose, which enabled us to prosecute our work. This line was more difficult than the first one; the glacier was broken into sharp-edged chasms; the ridges to be climbed were steep, and the snow which filled the depressions profound. The oblique arrangement of the crevasses also magnified the labour by increasing the circuits. I saw the leader of the party often shoulder-deep in snow, treading the soft mass as a swimmer walks in water, and I felt a wish to be at his side to cheer him and to share his toil. Each man there, however, knew my willingness to do this if occasion required it, and wrought contented. At length the last stake being fixed, the faces of the men were turned homeward. The evening became wilder, and the storm rose at times to a hurricane. On the more level portions of the glacier the snow lay deep and unsheltered; among its frozen waves and upon its more dislocated portions it had been partially engulfed, and the residue was more or less in shelter. Over the former spaces dense clouds of snow rose, whirling in the air and cutting off all view of the glacier. The whole length of the Mer de Glace was thus divided into clear and cloudy segments, and presented an aspect of wild and wonderful turmoil. A large pine stood near me, with its lowest branch spread out upon the surface of the snow; on this branch I seated myself, and, sheltered by the trunk, waited until I saw my men in safety. The wind caught the

branches of the trees, shook down their loads of snow,
and tossed it wildly in the air. Every mountain gave
a quota to the storm. The scene was one of most impressive grandeur, and the moan of the adjacent pines
chimed in noble harmony with the picture which addressed
the eyes.

At length we all found ourselves in safety within
doors. The windows shook violently. The tempest was
however intermittent throughout, as if at each effort it
had exhausted itself, and required time to recover its
strength. As I heard its heralding roar in the gullies of
the mountains, and its subsequent onset against our habitation, I thought wistfully of my stations, not knowing
whether they would be able to retain their positions in
the face of such a blast. That night however, as if the
storm had sung our lullaby, we all slept profoundly,
having arranged to commence our measurements as early
as light permitted on the following day.

Thursday, 29th December.—"Snow, heavy snow: it must
have descended throughout the entire night; the quantity
freshly fallen is so great; the atmosphere at seven o'clock
is thick with the descending flakes." At eight o'clock it
cleared up a little, and I proceeded to my station, while
the men advanced upon the glacier; but I had scarcely
fixed my theodolite when the storm recommenced. I
had a man to clear away the snow and otherwise assist me;
he procured an old door from the hotel, and by rearing
it upon its end sheltered the object-glass of the instrument. Added to the flakes descending from the clouds
was the spitting snow-dust raised by the wind, which
for a time so blinded me that I was unable to see the glacier. The measurement of the first stake was very
tedious, but practice afterwards enabled me to take advantage of the brief lulls and periods of partial clearness
with which the storm was interfused.

At nine o'clock my telescope happened to be directed upon the men as they struggled through the snow; all evidence of the deep track which they had formed yesterday having been swept away. I saw the leader sink and suddenly disappear. He had stood over a concealed fissure, the roof of which had given way and he had dropped in. I observed a rapid movement on the part of the remaining three men: they grouped themselves beside the fissure, and in a moment the missing man was drawn from between its jaws. His disappearance and appearance were both extraordinary. We had, as I have stated, provided for contingencies of this kind, and the man's rescue was almost immediate.

My attendant brought two poles from the hotel which we thrust obliquely into the snow, causing the free ends to cross each other; over these a blanket was thrown, behind which I sheltered myself from the storm as the men proceeded from stake to stake. At 9.30 the storm was so thick that I was unable to see the men at the stake which they had reached at the time; the flakes sped wildly in their oblique course across the field of the telescope. Some time afterwards the air became quite still, and the snow underwent a wonderful change. Frozen flowers similar to those I had observed on Monte Rosa fell in myriads. For a long time the flakes were wholly composed of these exquisite blossoms entangled together. On the surface of my woollen dress they were soft as down; the snow itself on which they fell seemed covered by a layer of down; while my coat was completely spangled with six-rayed stars. And thus prodigal Nature rained down beauty, and had done so here for ages unseen by man. And yet some flatter themselves with the idea that this world was planned with strict reference to human use; that the lilies of the field exist simply to appeal to the sense of the beautiful in man. True, this result is secured, but it

is one of a thousand all equally important in the eyes of
Nature. Whence those frozen blossoms? Why for Æons
wasted? The question reminds one of the poet's answer
when asked whence was the Rhodora:—

> "Why wert thou there, O rival of the rose?
> I never thought to ask, I never knew;
> But in my simple ignorance suppose
> The selfsame power that brought me there brought you!"*

I sketched some of the crystals, but, instead of repro-
ducing these sketches, which were rough and hasty, I have
annexed two of the forms drawn with so much skill and
patience by Mr. Glaisher.

We completed the measurement of the first line before
eleven o'clock, and I felt great satisfaction in the thought
that I possessed something of which the weather could not
deprive me. As I closed my note-book and shifted the
instrument to the second station, I felt that my expedition was already a success.

At a quarter past eleven I had my theodolite again fixed,
and ranging the telescope along the line of pickets, I saw
them all standing. Crossing the ice wilderness, and suggesting the operation of intelligence amid that scene of
desolation, their appearance was pleasant to me. Just
before I commenced, a solitary jay perched upon the summit of an adjacent pine and watched me. The air was
still at the time, and the snow fell heavily. The flowers
moreover were magnificent, varying from about the twentieth of an inch to two lines in diameter, while, falling
through the quiet air, their forms were perfect. Adjacent
to my theodolite was a stump of pine, from which I had
the snow removed, in order to have something to kick my
toes against when they became cold; and on the stump
was placed a blanket to be used as a screen in case of need.
While I remained at the station a layer of snow an inch

* Emerson.

Fig. 14.

Fig. 15.

thick fell upon this blanket, the whole layer being composed of these exquisite flowers. The atmosphere also was filled with them. From the clouds to the earth Nature was busy marshalling her atoms, and putting to shame by the beauty of her structures the comparative barbarities of Art.

My men at length reached the first station, and the measurement commenced. The storm drifted up the valley, thickening all the air as it approached. Denser and denser the flakes fell; but still, with care and tact I was able to follow my party to a distance of 800 yards. I had not thought it possible to see so far through so dense a storm. At this distance also my voice could be heard, and my instructions understood; for once, as the man who took up the line stood behind his baton and prevented its projection against the white snow, I called out to him to stand aside, and he promptly did so. Throughout the entire measurement the snow never ceased falling, and some of the illusions which it produced were extremely singular. The distant boundary of the glacier appeared to rise to an extraordinary height, and the men wading through the snow appeared as if climbing up a wall. The labour along this line was still greater than on the former; on the steeper slopes especially the toil was great; for here the effort of the leader to lift his own body added itself to that of cutting his way through the snow. His footing I could see often yielded, and he slid back, checking his recession, however, by still plunging forward; thus, though the limbs were incessantly exerted, it was, for a time, a mere motion of vibration without any sensible translation. At the last stake the men shouted, "*Nous sommes finis!*" and I distinctly heard them through the falling snow. By this time I was quite covered with the crystals which clung to my wrapper. They also formed a heap upon my theodolite, rising over the spirit-levels and embracing the lower portion of the vertical arc. The work was done; I struck

my theodolite and ascended to the hotel; the greatest
depth of snow through which I waded reaching, when I
stood erect, to within three inches of my breast.

The men returned; dinner was prepared and consumed;
the disorder which we had created made good; the rooms
were swept, the mattresses replaced, and the shutters fast-
ened, where this was possible. We locked up the house,
and with light hearts and lithe limbs commenced the descent. My aim now was to reach the source of the Arveiron, to examine the water and inspect the vault. With
this view we went straight down the mountain. The inclinations were often extremely steep, and down these we swept
with an avalanche-velocity; indeed usually accompanied
by an avalanche of our own creation. On one occasion
Balmat was for a moment overwhelmed by the descending
mass: the guides were startled, but he emerged instantly.
Tairraz followed him, and I followed Tairraz, all of us rolling in the snow at the bottom of the slope as if it were so
much flour. My practice on the Finsteraarhorn rendered
me at home here. One of the porters could by no means
be induced to try this flying mode of descent. Simond
carried my theodolite box, tied upon a crotchet on his
back; and once, while shooting down a slope, he incau-
tiously allowed a foot to get entangled; his momentum
rolled him over and over down the incline, the theodolite
emerging periodically from the snow during his successive
revolutions. A succession of *glissades* brought us with
amazing celerity to the bottom of the mountain, whence
we picked our way amid the covered boulders and over the
concealed arms of the stream to the source of the Arveiron.

The quantity of water issuing from the vault was considerable, and its character that of true glacier water. It was turbid
with suspended matter, though not so turbid as in summer;
but the difference in force and quantity would, I think, be
sufficient to account for the greater summer turbidity. This

character of the water could only be due to the grinding motion of the glacier upon its bed; a motion which seems not to be suspended even in the depth of winter. The temperature of the water was the tenth of a degree Centigrade above zero; that of the ice was half a degree below zero: this was also the temperature of the air, while that of the snow, which in some places covered the ice-blocks, was a degree and a quarter below zero.

The entrance to the vault was formed by an arch of ice which had detached itself from the general mass of the glacier behind: between them was a space through which we could look to the sky above. Beyond this the cave narrowed, and we found ourselves steeped in the blue light of the ice. The roof of the inner arch was perforated at one place by a shaft about a yard wide, which ran vertically to the surface of the glacier. Water had run down the sides of this shaft, and, being re-frozen below, formed a composite pillar of icicles at least twenty feet high and a yard thick, stretching quite from roof to floor. They were all united to a common surface at one side, but at the other they formed a series of flutings of exceeding beauty. This group of columns was bent at its base as if it had yielded to the forward motion of the glacier, or to the weight of the arch overhead. Passing over a number of large ice blocks which partially filled the interior of the vault, we reached its extremity, and here found a sloping passage with a perfect arch of crystal overhead, and leading by a steep gradient to the air above. This singular gallery was about seventy feet long, and was floored with snow. We crept up it, and from the summit descended by a glissade to the frontal portion of the cavern. To me this crystal cave, with the blue light glistening from its walls, presented an aspect of magical beauty. My delight, however, was tame compared with that of my companions.

Looking from the blue arch westwards, the heavens were

L

seen filled by crimson clouds, with fiery outlyers reaching up to the zenith. On quitting the vault I turned to have a last look at those noble sentinels of the Mer de Glace, the Aiguille du Dru, and the Aiguille Verte. The glacier below the mountains was in shadow, and its frozen precipices of a deep cold blue. From this, as from a basis, the mountain cones sprang steeply heavenward, meeting half way down the fiery light of the sinking sun. The right-hand slopes and edges of both pyramids burned in this light, while detached protuberant masses also caught the blaze, and mottled the mountains with effulgent spaces. A range of minor peaks ran slanting downwards from the summit of the Aiguille Verte ; some of these were covered with snow, and shone as if illuminated with the deep crimson of a strontian flame. I was absolutely struck dumb by the extraordinary majesty of this scene, and watched it silently till the red light faded from the highest summits. Thus ended my winter expedition to the Mer de Glace.

Next morning, starting at three o'clock, I was driven by my two guides in an open sledge to Sallenches. The rain was pitiless and the road abominable. The distance, I believe, is only six leagues, but it took us five hours to accomplish it. The leading mule was beyond the reach of Simond's whip, and proved a mere obstructive; during part of the way it was unloosed, tied to the sledge, and dragged after it. Simond afterwards mounted the hind-most beast and brought his whip to bear upon the leader; the jerking he endured for an hour and a half seemed almost sufficient to dislocate his bones. We reached Sallenches half an hour late, but the diligence was behind its time by this exact interval. We met it on the Pont St. Martin, and I transferred myself from the sledge to the interior. This was the morning of the 30th of December, and on the evening of the 1st of January I was in London.

I cannot finish this recital without saying one word about my men. Their behaviour was admirable throughout. The labour was enormous, but it was manfully and cheerfully done. I know Simond well; he is intelligent, truthful, and affectionate, and there is no guide of my acquaintance for whom I have a stronger regard. Joseph Tairraz is an extremely intelligent and able guide, and on this trying occasion proved himself worthy of my highest praise and commendation. Their two companions upon the glacier, Edouard Balmat (le Petit Balmat) and Joseph Simond (fils d'Auguste) acquitted themselves admirably; and it also gives me pleasure to bear testimony to the willing and efficient service of Francois Ravanal, who attended upon me during the observations.

PART II.

CHIEFLY SCIENTIFIC.

~~~~~~~~~~~~~~~~

Aber im stillen Gemach entwirft bedeutende Zirkel
   Sinnend der Weise, beschleicht forschend den schaffenden Geist,
Prüft der Stoffe Gewalt, der Magnete Hassen und Lieben,
   Folgt durch die Lüfte dem Klang, folgt durch den Aether dem Strahl,
Sucht das vertraute Gesetz in des Zufall's grausenden Wundern,
   Sucht den ruhenden Pol in der Erscheinungen Flucht.

<div align="right">SCHILLER.</div>

ON LIGHT AND HEAT.

(1.)

WHAT is Light? The ancients supposed it to be something emitted by the eyes, and for ages no notion was
entertained that it required time to pass through space.
In the year 1676 Römer first proved that the light from
Jupiter's satellites required a certain time to cross the
earth's orbit. Bradley afterwards found that, owing to
the velocity with which the earth flies through space,
the rays of the stars are slightly inclined, just as rain-
drops which descend vertically appear to meet us when
we move swiftly through the shower. In Kew Gardens
there is a sun-dial commemorative of this discovery, which
is called the *aberration of light*. Knowing the velocity of
the earth, and the inclination of the stellar rays, Bradley
was able to calculate the velocity of light; and his result
agrees closely with that of Römer. Celestial distances
were here involved, but a few years ago M. Fizeau, by
an extremely ingenious contrivance, determined the time
required by light to pass over a distance of about 9000
yards; and his experiment is quite in accordance with the
results of his predecessors.

But what is it which thus moves? Some, and among
the number Newton, imagined light to consist of particles
darted out from luminous bodies. This is the so-called
Emission-Theory, which was held by some of the greatest
men: Laplace, for example, accepted it; and M. Biot
has developed it with a lucidity and power peculiar to
himself. It was first opposed by the astronomer Huyghens,
and afterwards by Euler, both of whom supposed light to
be a kind of undulatory motion; but they were borne

down by their great antagonists, and the emission-theory held its ground until the commencement of the present century, when Thomas Young, Professor of Natural Philosophy in the Royal Institution, reversed the scientific creed by placing the Theory of Undulation on firm foundations. He was followed by a young Frenchman of extraordinary genius, who, by the force of his logic and the conclusiveness of his experiments, left the Wave-Theory without a competitor. The name of this young Frenchman was Augustin Fresnel.

Since his time some of the ablest minds in Europe have been applied to the investigation of this subject; and thus a mastery, almost miraculous, has been attained over the grandest and most subtle of natural phenomena. True knowledge is always fruitful, and a clear conception regarding any one natural agent leads infallibly to better notions regarding others. Thus it is that our knowledge of light has corrected and expanded our knowledge of *heat*, while the latter, in its turn, will assuredly lead us to clearer conceptions regarding the other forces of Nature.

I think it will not be a useless labour if I here endeavour to state, in a simple manner, our present views of light and heat. Such knowledge is essential to the explanation of many of the phenomena referred to in the foregoing pages; and even to the full comprehension of the origin of the glaciers themselves. A few remarks on the nature of sound will form a fit introduction.

It is known that sound is conveyed to our organs of hearing by the air: a bell struck in a vacuum emits no sound, and even when the air is thin the sound is enfeebled. Hawksbee proved this by the air-pump; De Saussure fired a pistol at the top of Mont Blanc,—I have repeated the experiment myself, and found, with him, that the sound is feebler than at the sea level. Sound is not produced by anything projected through the air. The explosion of a gun, for example, is sent forward by a motion of a totally different kind from that which animates the bullet projected from the gun: the latter is a motion of *translation;* the former, one of *vibration.* To use a rough comparison. sound is projected through the air as a push is through a crowd; it is the propagation of a *wave* or *pulse,* each particle taking up the motion of its neighbour, and delivering it on to the next. These aërial waves enter the external ear, meet a membrane, the so-called tympanic membrane, which is drawn across the passage at a certain place, and break upon it as sea-waves do upon the shore. The membrane is shaken, its tremors are communicated to the auditory nerve, and transmitted by it to the brain, where they produce the impression to which we give the name of sound.

In the tumult of a city, pulses of different kinds strike irregularly upon the tympanum, and we call the effect *noise;* but when a succession of impulses reach the ear *at regular intervals* we feel the effect as *music.* Thus, a vibrating string imparts a series of shocks to the air around it, which are transmitted with perfect regularity to the ear, and produce *a musical note.* When we hear the song of a soaring lark we may be sure that the entire atmosphere between us and the bird is filled with pulses, or undulations, or waves, as they are often called, produced by the little songster's organ of voice. This organ is a vibrating instrument, resembling, in principle, the reed of a clarionet. Let us suppose that we hear the song of a lark, elevated to a height of 500 feet in the air. Before this is possible, the bird must have agitated a sphere of air 1000 feet in diameter; that is to say, it must have communicated to 17,888 tons of air a motion sufficiently intense to be appreciated by our organs of hearing.

Musical sounds differ in *pitch:* some notes are high and shrill, others low and deep. Boys are chosen as choristers

to produce the shrill notes; men are chosen to produce
the bass notes.   Now, the sole difference here is, that the
boy's organ vibrates *more rapidly* than the man's—it sends
a greater number of impulses per second to the ear.
In like manner, a short string emits a higher note than a
long one, because it vibrates more quickly.   The greater
the number of vibrations which any instrument performs
in a given time, the higher will be the pitch of the note
produced.   The reason why the hum of a gnat is shriller
than that of a beetle is that the wings of the small
insect vibrate more quickly than those of the larger one.
We can, with suitable arrangements, make those sonorous
vibrations visible to the eye;* and we also possess instruments which enable us to tell, with the utmost exacti-
tude, the number of vibrations due to any particular note.
By such instruments we learn that a gnat can execute
many thousand flaps of its little wings in a second of time.

   In the study of nature the coarser phenomena, which
come under the cognizance of the senses, often suggest to
us the finer phenomena which come under the cognizance
of the mind; and thus the vibrations which produce sound,
and which, as has been stated, can be rendered visible to
the eye by proper means, first suggested that *light* might
be due to a somewhat similar action.   This is now the
universal belief.   A luminous body is supposed to have its
atoms, or molecules, in a state of intense vibration.   The

---

   * The vibrations of the air of a room in which a musical instrument is
sounded may be made manifest by the way in which fine sand arranges
itself upon a thin stretched membrane over which it is strewn; and indeed
Savart has thus rendered visible the vibrations of the tympanum itself.
Every trace of sand was swept from a paper drum held in the clock-tower
of Westminster when the Great Bell was sounded. Another way of showing
the propagation of aërial pulses is to insert a small gas jet into a vertical
glass tube about a foot in length, in which the flame may be caused to
burn tranquilly.   On pitching the voice to the note of an open tube a
foot long, the little flame quivers, stretches itself, and responds by producing a clear melodious note of the same pitch as that which excited
it.   The flame will continue its song for hours without intermission.

motions of the atoms are supposed to be communicated to a medium suited to their transmission, as air is to the transmission of sound. This medium is called the *luminiferous ether*, and the little billows excited in it speed through it with amazing celerity, enter the pupil of the eye, pass through the humours, and break upon the retina or optic nerve, which is spread out at the back of the eye. Hence the tremors they produce are transmitted along the nerve to the brain, where they announce themselves as *light*. The swiftness with which the waves of light are propagated through the ether, is however enormously greater than that with which the waves of sound pass through the air. An aërial wave of sound travels at about the rate of 1100 feet in a second: a wave of light leaves 192,000 miles behind it in the same time.

Thus, then, in the case of sound, we have the sonorous body, the air, and the auricular nerve, concerned in the phenomenon; in the case of light, we have the luminous body, the ether, and the optic nerve. The fundamental analogy of sound and light is thus before us, and it is easily remembered. But we must push the analogy further. We know that the white light which comes to us from the sun is made up of an infinite number of coloured rays. By refraction with a prism we can separate those rays from each other, and arrange them in the series of colours which constitute the solar spectrum. The rainbow is an imperfect or *impure* spectrum, produced by the drops of falling rain, but by prisms we can unravel the white light into pure red, orange, yellow, green, blue, indigo, and violet. Now, this spectrum is to the eye what the gamut is to the ear; each colour represents a note, and *the different colours represent notes of different pitch*. The vibrations which produce the impression of red are *slower*, and the waves which they produce are *longer*, than those to which we owe the sensation of violet; while the

vibrations which excite the other colours are intermediate
between these two extremes. This, then, is the second
grand analogy between light and sound : *Colour answers to
Pitch*. There is therefore truth in the figure when we say
that the gentian of the Alps sings a shriller note than the
wild rhododendron, and that the red glow of the mountains
at sunset is of a lower pitch than the blue of the firmament
at noon.

These are not fanciful analogies. To the mind of the
philosopher these waves of ether are almost as palpable
and certain as the waves of the sea, or the ripples on the
surface of a lake. The length of the waves, both of sound
and light, and the number of shocks which they respectively impart to the ear and eye, have been the subjects of
the strictest measurement. Let us here go through a
simple calculation. It has been found that 39,000 waves
of red light placed end to end would make up an inch.
How many inches are there in 192,000 miles? My
youngest reader can make the calculation for himself,
and find the answer to be 12,165,120,000 inches. It is
evident that, if we multiply this number by 39,000, we
shall obtain the number of waves of red light in 192,000
miles; this number is 474,439,680,000,000. *All these
waves enter the eye in one second;* thus the expression
"I see red colour," strictly means, "My eye is now in
receipt of four hundred and seventy-four millions of
millions of impulses per second." To produce the impression of violet light a still greater number of impulses
is necessary; the wave-length of violet is the $\frac{1}{57500}$th
part of an inch, and the number of shocks imparted in a
second by waves of this length is, in round numbers, six
hundred and ninety-nine millions of millions. The other
colours of the spectrum, as already stated, rise gradually
in pitch from the red to the violet.

A very curious analogy between the eye and ear may

here be noticed. The range of seeing is different in different persons—some see a longer spectrum than others; that is to say, rays which are obscure to some are luminous to others. Dr. Wollaston pointed out a similar fact as regards hearing; the range of which differs in different individuals. Savart has shown that a good ear can hear a musical note produced by 8 shocks in a second; it can also hear a note produced by 24,000 shocks in a second; but there are ears in which the range is much more limited. It is possible indeed to produce a sound which shall be painfully shrill to one person, while it is quite unheard by another. I once crossed a Swiss mountain in company with a friend; a donkey was in advance of us, and the dull tramp of the animal was plainly heard by my companion; but to me this sound was almost masked by the shrill chirruping of innumerable insects which thronged the adjacent grass; my friend heard nothing of this, it lay quite beyond his range of hearing.

A third and most important analogy between sound and light is now to be noted; and it will be best understood by reference to something more tangible than either. When a stone is thrown into calm water a series of rings spread themselves around the centre of disturbance. If a second stone be thrown in at some distance from the first, the rings emanating from both centres will cross each other, and at those points where the ridge of one wave coincides with the ridge of another the water will be lifted to a greater height. At those points, on the contrary, where the ridge of one wave crosses the furrow of another, we have both obliterated, and the water restored to its ordinary level. Where two ridges or two furrows unite, we have a case of *coincidence;* but where a ridge and a furrow unite we have what is called *interference.* It is quite possible to send two systems of waves into the same channel, and to *hold back* one system a little, so that its ridges shall coincide with the furrows of the other system. The "interference" would be here complete, and the waves thus circumstanced would mutually destroy each other, smooth water being the result. In this way, by the addition of motion to motion, *rest* may be produced.

In a precisely similar manner two systems of sonorous waves can be caused to interfere and mutually to destroy each other: thus, by adding sound to sound, *silence* may be produced. Two beams of light also may be caused to interfere and effect their mutual extinction: thus, by adding light to light, we can produce *darkness*. Here indeed we have a critical analogy between sound and light—*the* one, in fact, which compels the most profound thinkers of the present day to assume that light, like sound, is a case of undulatory motion.

We see here the vision of the intellect prolonged beyond the boundaries of sense into the region of what might be considered mere imagination. But, unlike other imaginations, we can bring ours to the test of experiment; indeed, so great a mastery have we obtained over these waves, which eye has not seen, nor ear heard, that we can with mathematical certainty cause them to coincide or to interfere, to help each other or to destroy each other, at pleasure. It is perhaps possible to be a little more precise here. Let two stones—with a small distance between them—be dropped into water at the same moment; a system of circular waves will be formed round each stone. Let the distance from one little crest to the next following one be called *the length of the wave*, and now let us inquire what will take place at a point equally distant from the places where the two stones were dropped in. Fixing our attention upon the ridge of the first wave in each case, it is manifest that, as the water propagates both systems with the same velocity, the two foremost ridges will reach the point in question at the same moment; the ridge of one would therefore *coincide* with the ridge of the other, and the water at this point would be lifted to a height greater than that of either of the previous ridges.

Again, supposing that by any means we had it in our power to retard one system of waves so as to cause the first ridge of the one to be exactly one wave length *behind* the first ridge of the other, when they arrive at the point referred to. It is plain that the first ridge of the retarded system now falls in with the second ridge of the unretarded system, and we have another case of coincidence. A little reflection will shew the same to be true when one system is retarded any number of *whole wave-lengths;* the first ridge of the retarded system will always, at the point referred to, coincide with a *ridge* of the unretarded system.

But now suppose the one system to be retarded only *half a wave-length;* it is perfectly clear that in this case the first ridge of the retarded system would fall in with the first *furrow* of the unretarded system, and instead of coincidence we should have *interference.* One system, in fact, would tend to make a hollow at the point referred to, the other would tend to make a hill, and thus both systems would oppose and neutralise each other, so that neither the hollow nor the hill would be produced; the water would maintain its ordinary level. What is here said of a single half-wave-length of retardation, is also true if the retardation amount to any *odd* number of half-wave-lengths. In all such cases we should have the ridge of the one system falling in with the furrow of the other; a mutual destruction of the waves of both systems being the consequence. The same remarks apply when the point, instead of being equally distant from both stones, is an even or an odd number of semi-undulations farther from the one than from the other. In the former case we should have coincidence, and in the latter case interference, at the point in question.

To the eye of a person who understands these things, nothing can be more interesting than the rippling of water under certain circumstances. By the action of interference its surface is sometimes shivered into the most beautiful mosaic, shifting and trembling as if with a kind of visible music. When the tide advances over a sea-beach on a calm and sunny day, and its tiny ripples enter, at various points, the clear shallow pools which the preceding tide had left behind, the little wavelets run and climb and cross each other, and thus form a lovely *chasing*, which has its counterpart in the lines of light converged by the ripples upon the sand underneath. When waves are skilfully generated in a vessel of mercury, and a strong light reflected from the surface of the metal is received upon a screen, the most beautiful effects may be observed. The shape of the vessel determines, in part, the character of the figures produced ; in a circular dish of mercury, for example, a disturbance at the centre propagates itself in circular waves, which after reflection again encircle the centre. If the point of disturbance be a little removed from the centre, the intersections of the direct and reflected waves produce the magnificent chasing shown in the annexed figure (16), which I have borrowed from the excellent work on Waves by the Messrs. Weber. The luminous figure reflected from such a surface is exceedingly beautiful. When the mercury is lightly struck by a glass point, in a direction concentric with the circumference of the vessel, the lines of light run round the vessel in mazy coils, interlacing and unravelling themselves in the most wonderful manner. If the vessel be square, a splendid mosaic is produced by the crossing of the direct and reflected waves. Description, however, can give but a feeble idea of these exquisite effects ;—

"Thou canst not wave thy staff in the air,
    Or dip thy paddle in the lake,
But it carves the brow of beauty there,
    And the ripples in rhymes the oar forsake."

Now, all that we have said regarding the retardation of the waves of water, by a whole undulation and a semi-

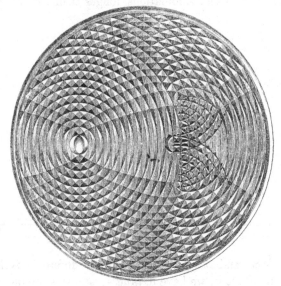

Fig. 16.

undulation, is perfectly applicable to the case of light. Two luminous points may be placed near to each other so as to resemble the two stones dropped into the water; and when the light of these is properly received upon a screen, or directly upon the retina, we find that at some places the action of the rays upon each other produces darkness, and at others augmented light. The former places are those where the rays emitted from one point are an *odd* number of semi-undulations in advance of the rays sent from the other; the latter places are those where the difference of path described by the rays is either nothing, or an *even* number of

semi-undulations. Supposing *a* and *b* (Fig. 17) to be two such
sources of light, and s r a screen on which the light falls;
at a point *l*, equally distant from *a* and *b*, we have *light*;
at a point *d*, where *a d* is half an undulation longer than
b d, we have darkness; at *l'*, where *a l'* is a whole wave-
length, or two semi-undulations, longer than *b l'*, we again

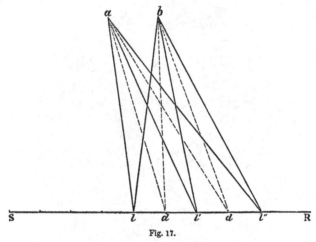

Fig. 17.

have light; and at a point *d'*,* where the difference is three
semi-undulations, we have darkness; and thus we obtain a
series of bright and dark spaces as we recede laterally
from the central point *l*.

Let a bit of tin foil be closely pasted upon a piece of glass,
and the edge of a penknife drawn across the foil so as to
produce a slit. Looking through this slit at a small and
distant light, we find the light spread out in a direction at
right angles to the slit, and if the light looked at be *mono-
chromatic*, that is, composed of a single colour, we shall
have a series of bright and dark bars corresponding to the
points at which the rays from the different points of the
slit alternately coincide and interfere upon the retina.

---

* The accent *'* is omitted in the figure; the second *d* counting from *l* is
meant.

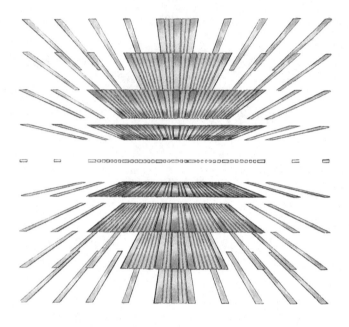

INTERFERENCE SPECTRA, PRODUCED BY DIFFRACTION.

FIG. 18.   *To face p.* 235.

By properly drawing a knife across a sheet of letter-paper a suitable slit may also be obtained; and those practised in such things can obtain the effect by looking through their fingers or their eyelashes.

But if the light looked at be white, the light of a candle for example, or of a jet of gas, instead of having a series of bright and dark bars, we have the bars *coloured*. And see how beautifully this harmonizes with what has been already said regarding the different lengths of the waves which produce different colours. Looking again at Fig. 17 we see that a certain obliquity is necessary to cause one ray to be a whole undulation in advance of the other at the point *l'*; but it is perfectly manifest that the obliquity must depend upon the length of the undulation; a long undulation would require a greater obliquity than a short one; red light, for example, requires a greater obliquity than blue light; so that if the point *l'* represents the place where the first bar of red light would be at its maximum strength, the maximum for blue would lie a little to the left of *l'*; the different colours are in this way separated from each other, and exhibit themselves as distinct fringes when a distant source of white light is regarded through a narrow slit.

By varying the shape of the aperture we alter the form of the chromatic image. A circular aperture, for example, placed in front of a telescope through which a point of white light is regarded, is seen surrounded by a concentric system of coloured rings. If we multiply our slits or apertures the phenomena augment in complexity and splendour. To give some notion of this I have copied from the excellent work of M. Schwerd the annexed figure (Fig. 18) which represents the gorgeous effect observed when a distant point of light is looked at through two gratings with slits of different widths.* A bird's feather repre-

* I am not aware whether in his own country, or in any other, a recognition at all commensurate with the value of the performance has followed

sents a peculiar system of slits, and the effect observed on properly looking through it is extremely interesting.

There are many ways by which the retardation necessary to the production of interference is effected. The splendid colours of a soap-bubble are entirely due to interference; the beam falling upon the transparent film is partially reflected at its outer surface, but a portion of it enters the film and is reflected at its *inner* surface. The latter portion having crossed the film and returned, is retarded, in comparison with the former, and, if the film be of suitable thickness, these two beams will clash and extinguish each other, while another thickness will cause the beams to coincide and illuminate the film with a light of greater intensity. From what has been said it must be manifest that to make two red beams thus coincide a thicker film would be required than would be necessary for two blue or green beams; thus, when the thickness of the bubble is suitable for the development of red, it is not suitable for the development of green, blue, &c.; the consequence is that we have different colours at different parts of the bubble. Owing to its compactness and to its being shaded by a covering of débris from the direct heat of the sun, the ice underneath the moraines of glaciers appears sometimes of a pitchy blackness. While cutting such ice with my axe I have often been surprised and delighted by sudden flashes of coloured light which broke like fire from the mass. These flashes were due to internal rupture, by which fissures were produced as thin as the film of a soap-bubble; the colours being due to the interference of the light reflected from the opposite sides of the fissures.

If spirit of turpentine, or olive oil, be thrown upon water, it speedily spreads in a thin film over the surface,

Schwerd's admirable essay entitled 'The Phenomena of Diffraction deduced from the Theory of Undulation.'

and the most gorgeous chromatic phenomena may be thus produced. Oil of lemons is also peculiarly suited to this experiment. If water be placed in a tea-tray, and light of sufficient intensity be suffered to fall upon it, this light will be reflected from the upper and under surfaces of the film of oil, and the colours thus produced may be received upon a screen, and seen at once by many hundred persons. If the oil of cinnamon be used, fine colours are also obtained, and the breaking up of this film exhibits a most interesting case of molecular action. By using a kind of varnish, instead of oil, Mr. Delarue has imparted such tenacity to these films that they may be removed from the water on which they rest and preserved for any length of time. By such films the colours of certain beetles, and of the wings of certain insects, may be accurately imitated; and a rook's feather may be made to shine with magnificent iridescences. The colours of tempered metals, and the beautiful metallochrome of Nobili are also due to a similar cause.

These colours are called the colours of *thin plates*, and are distinguished in treatises on optics from the coloured bars and fringes above referred to, and which are produced by *diffraction*, or the bending of the waves round the edge of an object. One result of this bending, which is of interest to us, was obtained by the celebrated Thomas Young. Permitting a beam of sunlight to enter a dark room through an aperture made with a fine needle, and placing in the path of the beam a bit of card one-thirtieth of an inch wide, he found the shadow of this card, or rather the line on which its shadow might be supposed to fall, always *bright;* and he proved the effect to be due to the bending of the waves of ether round the two edges of the card, and their coincidence at the other side. It has, indeed, been shown by M. Poisson, that the centre of the shadow of a small circular opaque disk which stands in

the way of a beam diverging from a point is exactly as
much illuminated as if the disk were absent. The singular effects described by M. Necker in the letter quoted
at page 178 at once suggest themselves here; and we see
how possible it is for the solar rays, in grazing a distant
tree, so to bend round it as to produce upon the retina,
where shadow might be expected, the impression of a tree
of light.* Another effect of diffraction is especially interesting to us at present. Let the seed of lycopodium
be scattered over a glass plate, or even like a cloud in
the air, and let a distant point of light be regarded through
it; the luminous point will appear surrounded by a series
of coloured rings, and when the light is intense, like the
electric or the Drummond light, the effect is exceedingly
fine.

And now for the application of these experiments. I
have already mentioned a series of coloured rings observed
around the sun by Mr. Huxley and myself from the Rhone
glacier; I have also referred to the cloud iridescences on
the Aletschhorn; and to the colours observed during my
second ascent of Monte Rosa, the magnificence of which
is neither to be rendered by pigments nor described in
words. All these splendid phenomena are, I believe, produced by diffraction, the vesicles or spherules of water in the
case of the cloud acting the part of the sporules in the case of
the lycopodium. The coloured fringe which surrounds the
Spirit of the Brocken, and the spectra which I have spoken
of as surrounding the sun, are also produced by diffraction.
By the interference of their rays in the earth's atmosphere the stars can momentarily quench themselves; and
probably to an intermittent action of this kind their twink-
ling, and the swift chromatic changes already mentioned,
are due. Does not all this sound more like a fairy

---

* I think, however, that the strong irradiation from the glistening sides
of the twigs and branches must also contribute to the result.

tale than the sober conclusions of science? What effort of
the imagination could transcend the realities here presented to us? The ancients had their spheral melodies,
but have not we ours, which only want a sense sufficiently
refined to hear them? Immensity is filled with this music; wherever a star sheds its light its notes are heard.
Our sun, for example, thrills concentric waves through
space, and every luminous point that gems our skies is
surrounded by a similar system. I have spoken of the
rising, climbing and crossing of the tiny ripples of a calm
tide upon a smooth strand; but what are they to those
intersecting ripples of the "uncontinented deep" by which
Infinity is engine-turned! Crossing solar and stellar distances, they bring us the light of sun and stars; thrilled
back from our atmosphere, they give us the blue radiance
of the sky; rounding liquid spherules, they clash at the
other side, and the survivors of the tumult bear to our
vision the wondrous cloud-dyes of Monte Rosa.

———————

(2.)

Thus, then, we have been led from Sound to Light, and
light now in its turn will lead us to *Radiant Heat;* for
in the order in which they are here mentioned the
conviction arose that they are all three different kinds of
motion. It has been said that the beams of the sun consist of rays of different colours, but this is not a complete
statement of the case. The sun emits a multitude of rays
which are perfectly non-luminous; and the same is true, in a
still greater degree, of our artificial sources of illumination.
Measured by the quantity of heat which they produce, 90
per cent. of the rays emanating from a flame of oil are

obscure ; while 99 out of every 100 of those which emanate from an alcohol flame are of the same description.*

In fact, the visible solar spectrum simply embraces an interval of rays of which the eye is formed to take cognizance, but it by no means marks the limits of solar action. Beyond the violet end of the spectrum we have obscure rays capable of producing chemical changes, and beyond the red we have rays possessing a high heating power, but incapable of exciting the impression of light. This latter fact was first established by Sir William Herschel, and it has been amply corroborated since.

The belief now universally prevalent is, that the rays of heat differ from the rays of light simply as one colour differs from another. As the waves which produce red are longer than those which produce yellow, so the waves which produce this obscure heat are longer than those which produce red. In fact, it may be shown that the longest waves never reach the retina at all ; they are completely absorbed by the humours of the eye.

What is true of the sun's obscure rays is also true of calorific rays emanating from any obscure source,—from our own bodies, for example, or from the surface of a vessel containing boiling water. We must, in fact, figure a warm body also as having its particles in a state of vibra-tion. When these motions are communicated from particle to particle of the body the heat is said to be *conducted ;* when, on the contrary, the particles transmit their vibra-tions through the surrounding ether, the heat is said to be *radiant.* This radiant heat, though obscure, exhibits a deportment exactly similar to light. It may be refracted and reflected, and collected in the focus of a mirror or of a suitable lens. The principle of interference also applies to it, so that by adding heat to heat we can produce *cold.* The identity indeed is complete throughout, and, recurring

* Melloni.

to the analogy of sound, we might define this radiant heat to be light of too low a pitch to be visible.

I have thus far spoken of *obscure* heat only; but the selfsame ray may excite both light and heat. The red rays of the spectrum possess a very high heating power. It was once supposed that the heat of the spectrum was an essence totally distinct from its light; but a profounder knowledge dispels this supposition, and leads us to infer that the selfsame ray, falling upon the nerves of feeling, excites heat, and falling upon the nerves of seeing, excites light. As the same electric current, if sent round a magnetic needle, along a wire, and across a conducting liquid, produces different physical effects, so also the same agent acting upon different organs of the body affects our consciousness differently.

(3.)

Heat has been defined in the foregoing section as a motion of the molecules or atoms of a body; but though the evidence in favour of this view is at present overwhelming, I do not ask the reader to accept it as a certainty, if he feels sceptically disposed. In this case, I would only ask him to accept it as a *symbol*. Regarded as a mere physical image, a kind of paper-currency of the mind, convertible, in due time, into the gold of truth, the hypothesis will be found exceedingly useful.

All known bodies possess more or less of this molecular motion, and all bodies are communicating it to the ether in which they are immersed. Ice possesses it. Ice before it melts attains a temperature of 32° Fahr., but the substance in winter often possesses a temperature far below 32°, so that in rising to 32° it is *warmed*. In experimenting

M

with ice I have often had occasion to cool it to 100° and more below the freezing point, and to warm it afterwards up to 32°.

If then we stand before a wall of ice, the wall radiates heat to us, and we also radiate heat to it; but the quantity which we radiate being greater than that which the ice radiates, we lose more than we gain, and are consequently chilled. If, on the contrary, we stand before a warm stove, a system of exchanges also takes place; but here the quantity we receive is in excess of the quantity lost, and we are warmed by the difference.

In like manner the earth radiates heat by day and by night into space, and against the sun, moon, and stars. By day, however, the quantity received is greater than the quantity lost, and the earth is warmed; by night the conditions are reversed; the earth radiates more heat than is sent to her by the moon and stars, and she is consequently cooled.

But here an important point is to be noted :—the earth receives the heat of the sun, moon, and stars, in great part as *luminous* heat, but she gives it out as *obscure* heat. I do not now speak of the heat *reflected* by the earth into space, as the light of the moon is to us; but of the heat which, after it has been absorbed by the earth, and has contributed to warm it, is radiated into space, as if the earth itself were its independent source. Thus we may properly say that the heat radiated from the earth is *different in quality* from that which she has received from the sun.

In one particular especially does this difference of quality show itself; besides being non-luminous, the heat radiated from the earth is more easily intercepted and absorbed by almost all transparent substances. A vast portion of the sun's rays, for example, can pass instantaneously through a thick sheet of water; gunpowder could easily be fired by the heat of the sun's rays con-

verged by passing through a thick water lens; the drops
upon leaves in greenhouses often act as lenses, and cause
the sun to burn the leaves upon which they rest. But
with regard to the rays of heat emanating from an obscure source, they are all absorbed by a layer of water
less than the 20th of an inch in thickness: water is
opaque to such rays, and cuts them off almost as effectually
as a metallic screen. The same is true of other liquids,
and also of many transparent solids.

Assuming the same to be true of gaseous bodies, that
they also intercept the obscure rays much more readily
than the luminous ones, it would follow that while the
sun's rays penetrate our atmosphere with freedom, the
change which they undergo in warming the earth deprives
them in a measure of this penetrating power. They can
reach the earth, but *they cannot get back;* thus the atmosphere acts the part of a ratchet-wheel in mechanics; it
allows of motion in one direction, but prevents it in the
other.

De Saussure, Fourier, M. Pouillet, and Mr. Hopkins have
developed this speculation, and drawn from it consequences
of the utmost importance; but it nevertheless rested
upon a basis of conjecture. Indeed some of the eminent
men above-named deemed its truth beyond the possibility
of experimental verification. Melloni showed that for a distance of 18 or 20 feet the absorption of obscure rays by
the atmosphere was absolutely inappreciable. Hence, the
total absorption being so small as to elude even Melloni's
delicate tests, it was reasonable to infer that *differences* of
absorption, if such existed at all, must be far beyond the
reach of the finest means which we could apply to detect
them.

This exclusion of one of the three states of material
aggregation from the region of experiment was, however,
by no means satisfactory; for our right to infer, from the

deportment of a solid or a liquid towards radiant heat, the deportment of a gas, is by no means evident. In both liquids and solids we have the molecules closely packed, and more or less chained by the force of cohesion; in gases, on the contrary, they are perfectly free, and widely separated. How do we know that the interception of radiant heat by liquids and solids may not be due to an arrangement and comparative rigidity of their parts, which gases do not at all share? The assumption which took no note of such a possibility seemed very insecure, and called for verification.

My interest in this question was augmented by the fact, that the assumption referred to lies, as will be seen, at the root of the glacier question. I therefore endeavoured to fill the gap, and to do for gases and vapours what had been already so ably done for liquids and solids by Melloni. I tried the methods heretofore pursued, and found them unavailing; oxygen, hydrogen, nitrogen, and atmospheric air, examined by such methods, showed no action upon radiant heat. Nature was dumb, but the question occurred, "Had she been addressed in the proper language?" If the experimentalist is convinced of this, he will rest content even with a negative; but the absence of this conviction is always a source of discomfort, and a stimulus to try again.

The principle of the method finally applied is all that can here be referred to; and it, I hope, will be quite intelligible. Two beams of heat, from two distinct sources, were allowed to fall upon the same instrument,* and to contend there for mastery. When both beams were perfectly equal, they completely neutralized each other's action; but when one of them was in any sensible degree stronger than the other, the predominance of the former was shown by the instrument. It was so arranged that

---

* The opposite faces of a thermo-electric pile.

one of the conflicting beams passed through a tube which could be exhausted of air, or filled with any gas; thus varying at pleasure the medium through which it passed. The question then was, supposing the two beams to be equal when the tube was filled with air, will the exhausting of the tube disturb the equality? The answer was affirmative; the instrument at once showed that a greater quantity of heat passed through the vacuum than through the air.

The experiment was so arranged that the effect thus produced was very large as measured by the indications of the instrument. But the action of the simple gases, oxygen, hydrogen, and nitrogen, was incomparably less than that produced by some of the compound gases, while these latter again differed widely from each other. Vapours exhibited differences of equal magnitude. The experiments indeed proved that gaseous bodies varied among themselves, as to their power of transmitting radiant heat, just as much as liquids and solids. It was in the highest degree interesting to observe how a gas or vapour of perfect transparency, as regards light, acted like an opaque screen upon the heat. To the eye, the gas within the tube might be as invisible as the air itself, while to the radiant heat it behaved like a cloud which it was almost impossible to penetrate.

Applying the same method, I have found that from the sun, from the electric light, or from the lime-light, a large amount of heat can be selected, which is unaffected not only by *air*, but by the most energetic gases that experiment has revealed to me; while this same heat, when it has its *quality* changed by being rendered obscure, is powerfully intercepted. Thus the bold and beautiful speculation above referred to has been made an experimental fact; the radiant heat of the sun does certainly pass through the atmosphere to the earth with greater facility than the radiant heat of the earth can escape into space.

It is probable that, were the earth unfurnished with this atmospheric swathing, its conditions of temperature would be such as to render it uninhabitable by man; and it is also probable that a suitable atmosphere enveloping the most distant planet might render it, as regards temperature, perfectly habitable. If the planet Neptune, for example, be surrounded by an atmosphere which permits the solar and stellar rays to pass towards the planet, but cuts off the escape of the warmth which they excite, it is easy to see that such an accumulation of heat may at length take place as to render the planet a comfortable habitation for beings constituted like ourselves.*

But let us not wander too far from our own concerns. Where radiant heat is allowed to fall upon an absorbing substance, a certain thickness of the latter is always necessary for the absorption. Supposing we place a thin film of glass before a source of heat, a certain per-centage of the heat will pass through the glass, and the remainder will be absorbed. Let the transmitted portion fall upon a second film similar to the first, a smaller percentage than before will be absorbed. A third plate would absorb still less, a fourth still less; and, after having passed through a sufficient number of layers, the heat would be so *sifted* that all the rays capable of being absorbed by glass would be abstracted from it. Suppose all these films to be placed together so as to form a single thick plate of glass, it is evident that the plate must act upon the heat which falls upon it, in such a manner that the major portion is absorbed *near the surface at which the heat enters.* This has been completely verified by experiment.

Applying this to the heat radiated from the earth, it is

* See a most interesting paper on this subject by Mr. Hopkins in the Cambridge 'Transactions,' May, 1856.

manifest that the greatest quantity of this heat will be absorbed by the lowest atmospheric strata. And here we find ourselves brought, by considerations apparently remote, face to face with the fact upon which the existence of all glaciers depends, namely, the comparative coldness of the upper regions of the atmosphere. The sun's rays can pass in a great measure through these regions without heating them; and the earth's rays, which they might absorb, hardly reach them at all, but are intercepted by the lower portions of the atmosphere.*

Another cause of the greater coldness of the higher atmosphere is the expansion of the denser air of the lower strata when it ascends. The dense air makes room for itself by pushing back the lighter and less elastic air which surrounds it: *it does work*, and, to perform this work, a certain amount of heat must be consumed. It is the consumption of this heat—its absolute annihilation as heat—that chills the expanded air, and to this action a share of the coldness of the higher atmosphere must undoubtedly be ascribed. A third cause of the difference of temperature is the large amount of heat communicated, *by way of contact*, to the air of the earth's surface; and a fourth and final cause is the loss endured by the highest strata through radiation into space.

* See M. Pouillet's important Memoir on Solar Radiation. Taylor's Scientific Memoirs, vol. iv. p. 44.

ORIGIN OF GLACIERS.

(4.)

HAVING thus accounted for the greater cold of the higher atmospheric regions, its consequences are next to be considered. One of these is, that clouds formed in the lower portions of the atmosphere, in warm and temperate latitudes, usually discharge themselves upon the earth as rain; while those formed in the higher regions discharge themselves upon the mountains as snow. The snow of the higher atmosphere is often melted to rain in passing through the warmer lower strata: nothing indeed is more common than to pass, in descending a mountain, from snow to rain; and I have already referred to a case of this kind. The appearance of the grassy and pine-clad alps, as seen from the valleys after a wet night, is often strikingly beautiful; the level at which the snow turned to rain being distinctly marked upon the slopes. Above this level the mountains are white, while below it they are green. The eye follows this *snow-line* with ease along the mountains, and when a sufficient extent of country is commanded its regularity is surprising.

The term "*snow-line*," however, which has been here applied to a local and temporary phenomenon, is commonly understood to mean something else. In the case just referred to it marked the place where the supply of solid matter from the upper atmospheric regions, during a single fall, was exactly equal to its consumption; but the term is usually understood to mean the line along which the quantity of snow which falls *annually* is melted, and no more. Below this line each year's snow is completely cleared away by the summer heat; above it a residual layer abides, which gradually augments in thickness from the snow-line upwards.

Here then we have a fresh layer laid on every year; and it is evident that, if this process continued without interruption, every mountain which rises above the snow-line must augment annually in height; the waters of the sea thus piled, in a solid form, upon the summits of the hills, would raise the latter to an indefinite elevation. But, as might be expected, the snow upon steep mountain-sides frequently slips and rolls down in avalanches into warmer regions, where it is reduced to water. A comparatively small quantity of the snow is, however, thus got rid of, and the great agent which Nature employs to relieve her overladen mountains is the glaciers.

Let us here avoid an error which may readily arise out the foregoing reflections. The principal region of clouds and rain and snow extends only to a limited distance upwards in the atmosphere; the highest regions contain very little moisture, and were our mountains sufficiently lofty to penetrate those regions, the quantity of snow falling upon their summits would be too trifling to resist the direct action of the solar rays. These would annually clear the summits to a certain level, and hence, were our mountains high enough, we should have a *superior*, as well as an inferior, snow-line; the region of *perpetual snow* would form a belt, below which, in summer, snowless valleys and plains would extend, and above which snowless summits would rise.

———

(5.)

At its origin then a glacier is snow—at its lower extremity it is ice. The blue blocks that arch the source of

M 3

the Arveiron were once powdery snow upon the slopes of the Col du Géant. Could our vision penetrate into the body of the glacier, we should find that the change from white to blue essentially consists in the gradual expulsion of the air which was originally entangled in the meshes of the fallen snow. Whiteness always results from the intimate and irregular mixture of air and a transparent solid; a crushed diamond would resemble snow ; if we pound the most transparent rock-salt into powder we have a substance as white as the whitest culinary salt; and the colourless glass vessel which holds the salt would also, if pounded, give a powder as white as the salt itself. It is a law of light that in passing from one substance to another possessing a different power of refraction, a portion of it is always reflected. Hence when light falls upon a transparent solid mixed with air, at each passage of the light from the air to the solid and from the solid to the air a portion of it is reflected; and, in the case of a powder, this reflection occurs so frequently that the passage of the light is practically cut off. Thus, from the mixture of two perfectly transparent substances, we obtain an opaque one; from the intimate mixture of air and water we obtain foam ; clouds owe their opacity to the same principle ; and the condensed steam of a locomotive casts a shadow upon the fields adjacent to the line, because the sunlight is wasted in echos at the innumerable limiting surfaces of water and air.

The snow which falls upon high mountain-eminences has often a temperature far below the freezing point of water. Such snow is *dry*, and if it always continued so the formation of a glacier from it would be impossible. The first action of the summer's sun is to raise the temperature of the superficial snow to 32°, and afterwards to melt it. The water thus formed percolates through the colder mass underneath, and this I take to be the first active agency in expelling the air entangled in the snow. But as the liquid trickles over the surfaces of granules colder than itself it is partially deposited in a solid form on these surfaces, thus augmenting the size of the granules, and cementing them together. When the mass thus formed is examined, the air within it is found as *round bubbles.* Now it is manifest that the air caught in the irregular interstices of the snow can have no tendency to assume this form so long as the snow remains solid; but the process to which I have referred—the saturation of the lower portions of the snow by the water produced by the melting of the superficial portions—enables the air to form itself into globules, and to give the ice of the *névé* its peculiar character. Thus we see that, though the sun cannot get directly at the deeper portions of the snow, by liquifying the upper layer he charges it with heat, and makes it his messenger to the cold subjacent mass.

The frost of the succeeding winter may, I think, or may not, according to circumstances, penetrate through this layer, and solidify the water which it still retains in its interstices. If the winter set in with clear frosty wea ther, the penetration will probably take place; but if heavy snow occur at the commencement of winter, thus throwing a protective covering over the *névé,* freezing to any great depth may be prevented. Mr. Huxley's idea seems to be quite within the range of possibility, that water-cells may be transmitted from the origin of the glacier to its end, retaining their contents always liquid.

It was formerly supposed, and is perhaps still supposed by many, that the snow of the mountains is converted into the ice of the glacier by the process of saturation and freezing just indicated. But the frozen layer would not yet resemble glacie ice; rit is only at the deeper portions of the *névé* that we find an approximation to the true ice of

the glacier. This brings us to the second great agent in
the process of glacification, namely, pressure. The ice of the
névé at 32° may be squeezed or crushed with extreme facility; and if the force be applied slowly and with caution,
the yielding of the mass may be made to resemble the
yielding of a plastic body. In the depths of the *névé*,
where each portion of the ice is surrounded by a resistant
mass, rude crushing is of course out of the question.
The layers underneath yield with extreme slowness to
the pressure of the mass above them; they are squeezed,
but not rudely fractured; and even should rude fracture
occur, the ice, as shall subsequently be shown, possesses
the power of restoring its own continuity. Thus, then,
the lower portions of the *névé* are removed by pressure
more and more from the condition of snow, the air-bubbles
which give to the *névé*-ice its whiteness are more and
more expelled, and this process, continued throughout the
entire glacier, finally brings the ice to that state of magnificent transparency which we find at the termination
of the glacier of Rosenlaui and elsewhere. This is all
capable of experimental proof. The Messrs. Schlagint-
weit compressed the snow of the *névé* to compact ice; and
I have myself frequently obtained slabs of ice from snow
in London.

COLOUR OF WATER AND ICE.

(6.)

THE sun is continually sending forth waves of different lengths, all of which travel with the same velocity through the ether. When these waves enter a prism of glass they are *retarded*, but in different degrees. The shorter waves suffer the greatest retardation, and in consequence of this are most deflected from their straight course. It is this property which enables us to separate one from the other in the solar spectrum, and this separation proves that the waves are by no means inextricably entangled with each other, but that they travel independently through space.

In consequence of this independence, the same body may intercept one system of waves while it allows another to pass : on this quality, indeed, depend all the phenomena of colour. A red glass, for example, is red because it is so constituted that it destroys the shorter waves which produce the other colours, and transmits only the waves which produce red. I may remark, however, that scarcely any glass is of a *pure* colour; along with the *predominant* waves, some of the other waves are permitted to pass. The colours of flowers are also very impure; in fact, to get pure colours we must resort to a delicate prismatic analysis of white light.

It has already been stated that a layer of water less than the twentieth of an inch in thickness suffices to stop and destroy all waves of radiant heat emanating from an obscure source. The longer waves of the obscure heat cannot get through water, and I find that all transparent compounds which contain *hydrogen* are peculiarly hostile to the longer undulations. It is, I think, the presence of this element in the humours of the eye which prevents the extra red rays of the solar spectrum from reaching the retina. It is interesting to observe that while bisulphide of carbon, chloride of phosphorus, and other liquids which contain no hydrogen, permit a large portion of the rays emanating from an iron or copper ball, at a heat below redness, to pass through them with facility, the same thickness of substances equally transparent, but which contain hydrogen, such as ether, alcohol, water, or the vitreous humour of the eye of an ox, completely intercepts these obscure rays. The same is true of solid bodies; a very slight thickness of those which contain hydrogen offers an impassable barrier to all rays emanating from a non-luminous source.* But the heat thus intercepted is by no means lost; its *radiant form* merely is destroyed. Its waves are shivered upon the particles of the body, but they impart *warmth* to it, while the heat which retains its radiant form contributes in no way to the warmth of the body through which it passes.

Water then absorbs all the extra red rays of the sun, and if the layer be thick enough *it invades the red rays themselves.* Thus the greater the distance the solar beams travel through pure water the more are they deprived of those components which lie at the red end of the spectrum. The consequence is, that the light finally transmitted by the water, and which gives to it its colour, is *blue.*

I find the following mode of examining the colour of water both satisfactory and convenient:—A tin tube, fifteen feet long and three inches in diameter, has its two ends stopped securely by pieces of colourless plate glass. It

---

* What is here stated regarding hydrogen is true of all the liquids and solids which have hitherto been examined,—but whether any exceptions occur, future experience must determine. It is only when in combination that it exhibits this impermeability to the obscure rays.

is placed in a horizontal position, and pure water is poured
into it through a small lateral pipe, until the liquid reaches
half way up the glasses at the ends; the tube then holds
a semi-cylinder of water and a semi-cylinder of air. A
white plate, or a sheet of white paper, well illuminated, is
then placed at a little distance from one end of the tube,
and is looked at through the tube. Two semicircular
spaces are then seen, one by the light which has passed
through the air, the other by the light which has passed
through the water; and their proximity furnishes a means of
comparison, which is absolutely necessary in experiments
of this kind. It is always found that, while the former
semicircle remains white, the latter one is vividly coloured.*

When the beam from an electric lamp is sent through this
tube, and a convex lens is placed at a suitable distance from
its most distant end, a magnified image of the coloured
and uncoloured semicircles may be projected upon a
screen. Tested thus, I have sometimes found, after rain,
the ordinary pipe-water of the Royal Institution quite
opaque; while, under other circumstances, I have found
the water of a clear green. The pump-water of the Institution thus examined exhibits a rich sherry colour, while
distilled water is blue-green.

The blueness of the Grotto of Capri is due to the fact
that the light which enters it has previously traversed a
great depth of clear water. According to Bunsen's account,
the *laugs*, or cisterns of hot water, in Iceland must be
extremely beautiful. The water contains silica in solution,
which, as the walls of the cistern arose, was deposited
upon them in fantastic incrustations. These, though white,
when looked at through the water appear of a lovely blue,
which deepens in tint as the vision plunges deeper into the
liquid.

* In my own experiments I have never yet been able to obtain a pure
blue, the nearest approach to it being a blue-green.

Ice is a crystal formed from this blue liquid, the colour of which it retains. Ice is the most opaque of transparent solids to radiant heat, as water is the most opaque of liquids. According to Melloni, a plate of ice one twenty-fifth of an inch thick, which permits the rays of light to pass without sensible absorption, cuts off 94 per cent. of the rays of heat issuing from a powerful oil lamp, 99½ per cent. of the rays issuing from incandescent platinum, and the whole of the rays issuing from an obscure source. The above numbers indicate how large a portion of the rays emitted by our artificial sources of light is obscure.

When the rays of light pass through a sufficient thickness of ice the longer waves are, as in the case of water, more and more absorbed, and the final colour of the substance is therefore blue. But when the ice is filled with minute air-bubbles, though we should loosely call it *white*, it may exhibit, even in small pieces, a delicate blue tint. This, I think, is due to the frequent interior reflection which takes place at the surfaces of the air-cells; so that the light which reaches the eye from the interior may, in consequence of its having been reflected hither and thither, really have passed through a considerable thickness of ice. The same remark, as we have already seen, applies to the delicate colour of the newly fallen snow.

COLOURS OF THE SKY.

(7.)

IN treating of the Colours of Thin Plates we found that a certain thickness was necessary to produce blue, while a greater thickness was necessary for red. With that wonderful power of generalization which belonged to him, Newton thus applies this apparently remote fact to the blue of the sky :—"The blue of the first order, though very faint and little, may possibly be the colour of some substances, and particularly the azure colour of the skies seems to be of this order. For all vapours, when they begin to condense and coalesce into small parcels, become first of that bigness whereby such an azure is reflected, before they can constitute clouds of other colours. And so, this being the first colour which vapours begin to reflect, it ought to be the colour of the finest and most transparent skies, in which vapours are not arrived at that grossness requisite to reflect other colours, as we find it is by experience."

M. Clausius has written a most interesting paper, in which he endeavours to show that the minute particles of water which are supposed by Newton to reflect the light, cannot be little globes entirely composed of water, but bladders or hollow spheres : the vapour must be in what is generally termed the *vesicular* state. He was followed by M. Brücke, whose experiments prove that the suspended particles may be so small that the reasoning of M. Clausius may not apply to them.

But why assume the existence of such particles at all? —why not assume that the colour of the air is blue, and renders the light of the sun blue, after the fashion of a blue glass or a solution of the sulphate of copper? I have already referred to the great variation which the colour of the firmament undergoes in the Alps, and have remarked that this seems to indicate that the blue depends upon some variable constituent of the atmosphere. Further, we find that the blue light of the sky is *reflected* light; and there must be something in the atmosphere capable of producing this reflection; but this thing, whatever it is, produces another effect which the blue glass or liquid is unable to produce. These *transmit* blue light, whereas, when the solar beams have traversed a great length of air, as in the morning or the evening, they are yellow, or orange, or even blood-red, according to the state of the atmosphere:—the transmitted light and the reflected light of the atmosphere are then totally different in colour.

Goethe, in his celebrated 'Farbenlehre,' gives a theory of the colour of the sky, and has illustrated it by a series of striking facts. He assumed two principles in the universe—Light and Darkness—and an intermediate stage of Turbidity. When the darkness is seen through a turbid medium on which the light falls, the medium appears blue; when the light itself is viewed through such a medium, it is yellow, or orange, or ruby-red. This he applies to the atmosphere, which sends us blue light, or red, according as the darkness of infinite space, or the bright surface of the sun, is regarded through it.

As a theory of colours Goethe's work is of no value, but the facts which he has brought forward in illustration of the action of turbid media are in the highest degree interesting. He refers to the blueness of distant mountains, of smoke, of the lower part of the flame of a candle (which if looked at with a white surface behind it completely disappears), of soapy water, and of the precipitates of various resins in water. One of his anecdotes in connexion with this subject is extremely curious and instructive. The portrait of a very dignified theologian having suffered from dirt, it was given to a painter to be cleaned. The clergyman was drawn in a dress of black velvet, over which the painter, in the first place, passed his sponge. To his astonishment the black velvet changed to the colour of blue plush, and completely altered the aspect of its wearer. Goethe was informed of the fact; the experiment was repeated in his presence, and he at once solved it by reference to his theory. The varnish of the picture when mixed with the water formed a turbid medium, and the black coat seen through it appeared blue; when the water evaporated the coat resumed its original aspect.

With regard to the real explanation of these effects, it may be shown, that, if a beam of white light be sent through a liquid which contains extremely minute particles in a state of suspension, the short waves are more copiously reflected by such particles than the long ones; blue, for example, is more copiously reflected than red. This may be shown by various fine precipitates, but the best is that of Brücke. We know that mastic and various resins are soluble in alcohol, and are precipitated when the solution is poured into water: *Eau de Cologne*, for example, produces a white precipitate when poured into water. If however this precipitate be sufficiently diluted, it gives the liquid a bluish colour by reflected light. Even when the precipitate is very thick and gross, and floats upon the liquid like a kind of curd, its under portions often exhibit a fine blue. To obtain particles of a proper size, Brücke recommends 1 gramme of colourless mastic to be dissolved in 87 grammes of alcohol, and dropped into a beaker of water, which is kept in a state of agitation. In this way a blue resembling that of the firmament may be produced. It is best seen when a black cloth is placed behind the glass; but in certain positions this blue liquid appears yellow; and these are the positions when the *transmitted* light reaches the eye. It is evident that this change of colour must necessarily exist; for the blue being partially withdrawn by more copious reflection, the transmitted light must partake more or less of the character of the complementary colour; though it does not follow that they should be exactly complementary to each other.

When a long tube is filled with clear water, the colour of the liquid, as before stated, shows itself by transmitted light. The effect is very interesting when a solution of mastic is permitted to drop into such a tube, and the fine precipitate to diffuse itself in the water. The blue-green of the liquid is first neutralized, and a yellow colour shows itself; on adding more of the solution the colour passes from yellow to orange, and from orange to blood-red. With a cell an inch and a half in width, containing water, into which the solution of mastic is suffered to drop, the same effect may be obtained. If the light of an electric lamp be caused to form a clear sunlike disk upon a white screen, the gradual change of this light by augmented precipitation into deep glowing red, resembling the colour of the sun when seen through fine London smoke, is exceedingly striking. Indeed the smoke acts, in some measure, the part of our finely-suspended matter.

By such means it is possible to imitate the phenomena of the firmament; we can produce its pure blue, and cause it to vary as in nature. The milkiness which steals over the heavens, and enables us to distinguish one cloudless day from another, can be produced with the greatest ease. The yellow, orange, and red light of the morning and evening can also be obtained: indeed the effects are so strikingly alike as to suggest a common origin—that the colours of the sky are due to minute particles diffused through the atmosphere. These particles are doubtless

the condensed vapour of water, and its variation in quality
and amount enables us to understand the variability of the
firmamental blue, and of the morning and the evening red.
Professor Forbes, moreover, has made the interesting observation that the steam of a locomotive, at a certain stage
of its condensation, is blue or red according as it is viewed
by reflected or transmitted light.

These considerations enable us to account for a number
of facts of common occurrence. Thin milk, when poured
upon a black surface, appears bluish. The milk is colourless; that is, its blueness is not due to *absorption,* but to a
separation of the light by the particles suspended in the
liquid. The juices of various plants owe their blueness to
the same cause; but perhaps the most curious illustration
is that presented by a blue eye. Here we have no true
colouring matter, no proper absorption; but we look
through a muddy medium at the black choroid coat within
the eye, and the medium appears blue.*

Is it not probable that this action of finely-divided
matter may have some influence on the colour of some of
the Swiss lakes—as that of Geneva for example? This
lake is simply an expansion of the river Rhone, which
rushes from the end of the Rhone glacier, as the Arveiron
does from the end of the Mer de Glace. Numerous other
streams join the Rhone right and left during its downward course; and these feeders, being almost wholly derived from glaciers, join the Rhone charged with the
finer matter which these in their motion have ground
from the rocks over which they have passed. But the
glaciers must grind the mass beneath them to particles
of all sizes, and I cannot help thinking that the finest of
them must remain suspended in the lake throughout its
entire length. Faraday has shown that a precipitate of
gold may require months to sink to the bottom of a

* Helmholtz, 'Das Sehen des Menschen.'

bottle not more than five inches high, and in all probability it would require *ages* of calm subsidence to bring *all* the particles which the Lake of Geneva contains to its bottom. It seems certainly worthy of examination whether such particles suspended in the water contribute to the production of that magnificent blue which has excited the admiration of all who have seen it under favourable circumstances.

THE MORAINES.

(8.)

THE surface of the glacier does not long retain the shining whiteness of the snow from which it is derived. It is flanked by mountains which are washed by rain, dislocated by frost, riven by lightning, traversed by avalanches, and swept by storms. The lighter débris is scattered by the winds far and wide over the glacier, sullying the purity of its surface. Loose shingle rattles at intervals down the sides of the mountains, and falls upon the ice where it touches the rocks. Large rocks are continually let loose, which come jumping from ledge to ledge, the cohesion of some being proof against the shocks which they experience; while others, when they hit the rocks, burst like bomb-shells, and shower their fragments upon the ice.

Thus the glacier is incessantly loaded along its borders with the ruins of the mountains which limit it; and it is evident that the quantity of rock and rubbish thus cast upon the glacier depends upon the character of the adjacent mountains. Where the summits are bare and friable, we may expect copious showers; where they are resistant, and particularly where they are protected by a covering of ice and snow, the quantity will be small. As the glacier moves downward, it carries with it the load deposited upon it. Long ridges of débris thus flank the glacier, and these ridges are called *lateral moraines.* Where two tributary glaciers join to form a trunk-glacier, their adjacent lateral moraines are laid side by side at the place of confluence, thus constituting a ridge which runs along the middle of the trunk-glacier, and which is called a *medial moraine*. The rocks and débris carried down by the glacier are finally deposited at its lower extremity, forming there a *terminal moraine*.

It need hardly be stated that the number of medial moraines is only limited by the number of branch glaciers. If a glacier have but two branches, it will have only one medial moraine; if it have three branches, it will have two medial moraines; if n branches, it will have $n-1$ medial moraines. The number of medial moraines, in short, is always *one less* than the number of branches. A glance at the annexed figure will reveal the manner in which the lateral moraines of the Mer de Glace unite to form medial ones. (See Fig. 19.)

When a glacier diminishes in size it leaves its lateral moraines stranded on the flanks of the valleys. Successive shrinkings may thus occur, and *have* occurred at intervals of centuries; and a succession of old lateral moraines, such as many glacier-valleys exhibit, is the consequence. The Mer de Glace, for example, has its old lateral moraines, which run parallel with its present ones. The glacier may also diminish *in length* at distant intervals; the result being a succession of more or less concentric terminal moraines. In front of the Rhone-glacier we have six or seven such moraines, and the Mer de Glace also possesses a series of them.

Let us now consider the effect produced by a block of stone upon the surface of a glacier. The ice around it receives the direct rays of the sun, and is acted on by the warm air; it is therefore constantly melting. The stone also receives the solar beams, is warmed, and transmits its heat, by conduction, to the ice beneath it. If the heat thus transmitted to the ice through the stone be less than an equal space of the surrounding ice receives, it is manifest that the ice around the stone will waste more quickly than that beneath it, and the consequence is, that,

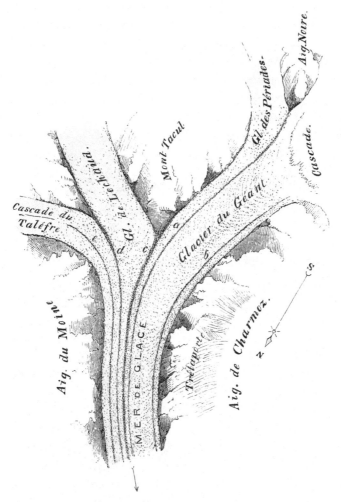

Cascade du Taléfre.

Gl. de Leckana.

Mont Tacul.

Gl. des Périades.

Aig. Noire.

Cascade.

Glacier du Géant.

Aig. du Moine.

MER DE GLACE.

Trélaporte.

Aig. de Charmoz.

MORAINES OF THE MER DE GLACE.

Fig. 19.   To face p. 264.

as the surface sinks, it leaves behind it a pillar of ice, on which the block is elevated. If the stone be wide and flat, it may rise to a considerable height, and in this position it constitutes what is called a glacier-*table*.

Almost all glaciers present examples of such tables; but no glacier with which I am acquainted exhibits them in greater number and perfection than the Unteraar glacier, near the Grimsel. Vast masses of granite are thus poised aloft on icy pedestals; but a limit is placed to their exaltation by the following circumstance. The sun plays obliquely upon the table all day; its southern extremity receives more heat than its northern, and the consequence is, that it *dips* towards the south. Strictly speaking, the plane of the dip rotates a little during the day, being a little inclined towards the east in the morning, north and south a little after noon, and inclined towards the west in the evening; so that, theoretically speaking, the block is a sun-dial, showing by its position the hour of the day. This rotation is, however, too small to be sensible, and hence *the dip of the stones upon a glacier sufficiently exposed to the sunlight, enables us at any time to draw the meridian line along its surface.* The inclination finally becomes so great that the block slips off its pedestal, and begins to form another, while the one which it originally occupied speedily disappears, under the influence of sun and air. Fig. 20 represents a typical section of a glacier-table, the sun's rays being supposed to fall in the direction of the shading lines.

Stones of a certain size are always lifted in the way described. A considerable portion of the heat which a large block receives is wasted by radiation, and by communication to the air, so that the quantity which reaches the ice beneath is trifling. Such a mass is, of course, a protector of the ice beneath it. But if the stone be small, and dark in colour, it absorbs the heat with avidity, com-

N

municates it quickly to the ice with which it is in contact, and consequently sinks in the ice. This is also the case with bits of dirt and the finer fragments of débris; they sink in the glacier. Sometimes, however, a pretty thick layer

Fig. 20.

of sand is washed over the ice from the moraines, or from the mountain-sides; and such sand-layers give birth to ice-cones, which grow to peculiarly grand dimensions on the Lower Aar glacier. I say "grow," but the truth, of course, is, that the surrounding ice wastes, while the portion underneath the sand is so protected that it remains as an eminence behind. At first sight, these sand-covered cones appear huge heaps of dirt, but on examination they are found to be cones of ice, and that the dirt constitutes merely a superficial covering.

Turn we now to the moraines. Protecting, as they do, the ice from waste, they rise, as might be expected, in vast ridges above the general surface of the glacier. In some cases the surrounding mass has been so wasted as to leave the spines of ice which support the moraines forty or fifty feet above the general level of the glacier. I should think the moraines

of the Mer de Glace about the Tacul rise to this height.
But lower down, in the neighbourhood of the Echelets,
these high ridges disappear, and nought remains to mark
the huge moraine but a strip of dirt, and perhaps a slight
longitudinal protuberance on the surface of the glacier.
How have the blocks vanished that once loaded the
moraines near the Tacul? They have been swallowed in
the crevasses which intersect the moraines lower down;
and if we could examine the ice at the Echelets we
should find the engulfed rocks in the body of the glacier.

Cases occur, wherein moraines, after having been
engulfed, and hidden for a time, are again entirely
disgorged by the glacier. Two moraines run along the
basin of the Talèfre, one from the Jardin, the other from
an adjacent promontory, proceeding parallel to each other
towards the summit of the great ice-fall. Here the ice is
riven, and profound chasms are formed, in which the
blocks and shingle of the moraines disappear. Throughout the entire ice-fall the only trace of the moraines is a
broad dirt-streak, which the eye may follow along the centre
of the fall, with perhaps here and there a stone which has
managed to rise from its frozen sepulchre. But the ice
wastes, and at the base of the fall large masses of stone
begin to reappear; these become more numerous as we
descend; the smaller débris also appears, and finally, at
some distance below the fall, the moraine is completely
restored, and begins to exercise its protecting influence;
it rises upon its ridge of ice, and dominates as before over
the surface of the glacier.

The ice under the moraines and sand-cones is of a
different appearance from that of the surrounding glacier,
and the principles we have laid down enable us to explain
the difference. The sun's rays, striking upon the unpro-
tected surface of the glacier, enter the ice to a considerable
depth; and the consequence is, that the ice near the

surface of the glacier is always disintegrated, being cut up with minute fissures and cavities, filled with water and air, which, for reasons already assigned, cause the glacier, when it is clean, to appear white and opaque. The ice under the moraines, on the contrary, is usually dark and transparent; I have sometimes seen it as black as pitch, the blackness being a proof of its great transparency, which prevents the reflection of light from its interior.

The ice under the moraines cannot be assailed in its depths by the solar heat, because this heat becomes *obscure* before it reaches the ice, and as such it lacks the power of penetrating the substance. It is also communicated in great part by way of contact instead of by radiation. A thin film at the surface of the moraine-ice engages all the heat that acts upon it, its deeper portions remaining intact and transparent.

GLACIER MOTION.

PRELIMINARY.

(9.)

THOUGH a glacier is really composed of two portions, one above and the other below the snow-line, the term glacier is usually restricted to the latter, while the French term *névé* is applied to the former. It is manifest that the snow which falls upon the glacier proper can contribute nothing to its growth or permanence; for every summer is not only competent to abolish the accumulations of the foregoing winter, but to do a great deal more. During each summer indeed a considerable quantity of the ice below the snow-line is reduced to water; so that, if the waste were not in some way supplied, it is manifest that in a few years the lower portion of the glacier must entirely disappear. The end of the Mer de Glace, for example, could never year after year thrust itself into the valley of Chamouni, were there not some agency by which its manifest waste is made good. This agency is the motion of the glacier.

To those unacquainted with the fact of their motion, but who have stood upon these vast accumulations of ice, and noticed their apparent fixity and rigidity, the assertion that a glacier moves must appear in the highest degree startling and incredible. They would naturally share the doubts of a certain professor of Tübingen, who, after a visit to the glaciers of Switzerland, went home and wrote a book flatly denying the possibility of their motion. But reflection comes to the aid of sense, and qualifies first impressions. We ask ourselves how is the permanence of the glacier secured? How are the moraines to be accounted for? Whence come the blocks which we often find at the terminus of a glacier, and which we know belong to distant mountains? The necessity of motion to produce these results becomes more and more apparent, until at length we resort to actual experiment. We take two fixed points at opposite sides of the glacier, so that a block of stone which rests upon the ice may be in the straight line which unites the points; and we soon find that the block quits the line, and is borne downwards by the glacier. We may well realize the interest of the man who first engaged in this experiment, and the pleasure which he felt on finding that the block moved; for even now, after hundreds of observations on the motion of glaciers have been made, the actual observance of this motion for the first time is always accompanied by a thrill of delight. Such pleasure the direct perception of natural truth always imparts. Like Antæus we touch our mother, and are refreshed by the contact.

The fact of glacier-motion has been known for an indefinite time to the inhabitants of the mountains; but the first who made quantitative observations of the motion was Hugi. He found that from 1827 to 1830 his cabin upon the glacier of the Aar had moved 100 mètres, or about 110 yards, downwards; in 1836 it had moved 714 mètres; and in 1841 M. Agassiz found it at a distance of 1428 mètres from its first position. This is equivalent in round numbers to an average velocity of 100 mètres a-year. In 1840 M. Agassiz fixed the position of the rock known as the Hôtel des Neufchâtelois; and on the 5th of September, 1841, he found that it had moved 213 feet downward. Between this date and September, 1842, the rock moved 273 feet, thus accomplishing a distance of 486 feet in two years.

But much uncertainty prevailed regarding the motion of the boulders, for they sometimes rolled upon the glacier,

and hence it was resolved to use stakes of wood driven into
the ice. In the month of July, 1841, M. Escher de la
Linth fixed a system of stakes, every two of which were
separated from each other by a distance of 100 mètres,
across the great Aletsch glacier. A considerable number
of other stakes were fixed *along* the glacier, the *longitudinal* separation being also 100 mètres. On the 8th of
July the stakes stood at a depth of about three feet in
the ice. On the 16th of August he returned to the glacier.
Almost all the stakes had fallen, and no trace, even of the
holes in which they had been sunk, remained. M. Agassiz
was equally unsuccessful on the glacier of the Aar. It
must therefore be borne in mind, that, previous to the
introduction of the facile modes of measurement which we
now employ, severe labour and frequent disappointment
had taught observers the true conditions of success.

After his defeat upon the Aletsch, M. Escher joined
MM. Agassiz and Desor on the Aar glacier, where, between
the 31st of August and the 5th of September, they fixed
in concert the positions of a series of blocks upon the ice,
with the view of measuring their displacements the following year.

Another observation of great importance was also commenced in 1841. Warned by previous failures, M. Agassiz
had iron boring-rods carried up the glacier, with which he
pierced the ice at six places to a depth of ten feet, and
at each place drove a wooden pile into the ice. These six
stations were in the same straight line across the glacier;
three of them standing upon the Finsteraar and three on
the Lauteraar tributary. About this time also M. Agassiz
conceived the idea of having the displacements measured
the year following with precise instruments, and also of
having constructed, by a professional engineer, a map of
the entire glacier, on which all its visible "accidents"
should be drawn according to scale. This excellent work

was afterwards executed by M. Wild, now Professor of Geodesy and Topography in the Polytechnic School of Zürich, and it is published as a separate atlas in connexion with M. Agassiz's 'Système Glaciaire.'

M. Agassiz is a naturalist, and he appears to have devoted but little attention to the study of physics. At all events, the physical portions of his writings appear to me to be very often defective. It was probably his own consciousness of this deficiency that led him to invoke the advice of Arago and others previous to setting out upon his excursions. It was also his desire "to see a philosopher so justly celebrated occupy himself with the subject," which induced him to invite Prof. J. D. Forbes of Edinburgh to be his guest upon the Aar glacier in 1841. On the 8th of August they met at the Grimsel Hospice, and for three weeks afterwards they were engaged together daily upon the ice, sharing at night the shelter of the same rude roof. It is in reference to this visit that Prof. Forbes writes thus at page 38 of the ' Travels in the Alps: '—"Far from being ready to admit, as my sanguine companions wished me to do in 1841, that the theory of glaciers was complete, and the cause of their motion certain, after patiently hearing all they had to say and reserving my opinion, I drew the conclusion that no theory which I had then heard of could account for the few facts admitted on all hands." In 1842 Prof. Forbes repaired, as early as the state of the snow permitted, to the Mer de Glace; he worked there, in the first instance, for a week, and afterwards crossed over to Courmayeur to witness a solar eclipse. The result of his week's observations was immediately communicated to Prof. Jameson, then editor of the ' Edinburgh New Philosophical Journal.'

In that letter he announces the fact, but gives no details of the measurement, that "the central part of the glacier moves faster than the edges in a very considerable proportion ; quite contrary to the opinion generally entertained." He also announced at the same time the continuous hourly advance of the glacier. This letter bears the date, " Courmayeur, Piedmont, 4th July," but it was not published until the month of October following.

Meanwhile M. Agassiz, in company with M. Wild, returned to complete his experiment upon the glacier of the Aar. On the 20th of July, 1842, the displacements of the six piles which he had planted the year before were determined by means of a theodolite. Of the three upon the Finsteraar affluent, that nearest the side had moved 160 feet, the next 225 feet, while that nearest to the centre had moved 269 feet. Of those on the Lauteraar, that nearest the side had moved 125 feet, the next 210 feet, and that nearest the centre 246 feet. These observations were perfectly conclusive as to the quicker motion of the centre: they embrace a year's motion; and the magnitude of the displacements, causing errors of inches, which might seriously affect small displacements, to vanish, justifies us in ranking this experiment with the most satisfactory of the kind that have ever been made. The results were communicated to Arago in a letter dated from the glacier of the Aar, on the 1st of August, 1842; they were laid before the Academy of Sciences on the 29th of August, 1842, and are published in the 'Comptes Rendus' of the same date.

The facts, then, so far as I have been able to collect them, are as follows:—M. Agassiz commenced his experiment about ten months before Professor Forbes, and the results of his measurements, with quantities stated, were communicated to the French Academy about two months prior to the publication of the letter of Professor Forbes in the 'Edinburgh Philosophical Journal.' But the latter communication, announcing in general terms the fact of the speedier central motion, was dated from

Courmayeur twenty-seven days before the date of M. Agassiz's letter from the glacier of the Aar.

The speedier motion of the central portion of a glacier has been justly regarded as one of cardinal importance, and no other observation has been the subject of such frequent reference; but the general impression in England is, that M. Agassiz had neither part nor lot in the establishment of the above fact; and in no English work with which I am acquainted can I find any reference to the above measurements. Relying indeed upon such sources for my information, I remained ignorant of the existence of the paper in the 'Comptes Rendus' until my attention was directed to it by Professor Wheatstone. In the next following chapters I shall have to state the results of some of my own measurements, and shall afterwards devote a little time to the consideration of the cause of glacier-motion. In treating a question on which so much has been written, it is of course impossible, as it would be undesireable, to avoid subjecting both my own views and those of others to a critical examination. But in so doing I hope that no expression shall escape me inconsistent with the courtesy which ought to be habitual among philosophers or with the frank recognition of the just claims of my predecessors.

MOTION OF THE MER DE GLACE.

(10.)

ON Tuesday, the 14th of July, 1857, I made my first observation on the motion of the Mer de Glace. Accompanied by Mr. Hirst I selected on the steep slope of the Glacier des Bois a straight pinnacle of ice, the front edge of which was perfectly vertical. In coincidence with this edge I fixed the vertical fibre of the theodolite, and permitted the instrument to stand for three hours. On looking through it at the end of this interval, the cross hairs were found projected against the white side of the pyramid; the whole mass having moved several inches downwards.

The instrument here mentioned, which had long been in use among engineers and surveyors, was first applied to measure glacier-motion in 1842; by Prof. Forbes on the Mer de Glace, and by M. Agassiz on the glacier of the Aar. The portion of the theodolite made use of is easily understood. The instrument is furnished with a telescope capable of turning up and down upon a pivot, without the slightest deviation right or left; and also capable of turning right or left without the slightest deviation up or down. Within the telescope two pieces of spider's thread, so fine as to be scarcely visible to the naked eye, are drawn across the tube and across each other. When we look through the telescope we see these fibres, their point of intersection being exactly in the centre of the tube; and the instrument is furnished with screws by means of which this point can be fixed upon any desired object with the utmost precision.

In setting a straight row of stakes across the glacier, our mode of proceeding was in all cases this:—The theodolite was placed on the mountain-side flanking the glacier, quite clear of the ice; and having determined the direction of a line perpendicular to the axis of the glacier, a well-defined object was sought at the opposite side of the valley as close as possible to this direction; the object being, in some cases, the sharp edge of a cliff; in others, a projecting corner of rock; and, in others, a well-defined mark on the face of the rock. This object and those around it were carefully sketched, so that on returning to the place it could be instantly recognized. On commencing a line the point of intersection of the two spiders' threads within the telescope was first fixed accurately upon the point thus chosen, and an assistant carrying a straight baton was sent upon the ice. By rough signalling he first stood near the place where the first stake was to be driven in; and the object end of the telescope was then lowered until he came within the field of view. He held his staff upright upon the ice, and, in obedience to signals, moved upwards or downwards until the point of intersection of the spiders' threads exactly hit the bottom of the baton; a concerted signal was then made, the ice was pierced with an auger to a depth of about sixteen inches, and a stake about two feet long was firmly driven into it. The assistant then advanced for some distance across the glacier; the end of the telescope was now gently raised until he and his upright staff again appeared in the field of view. He then moved as before until the bottom of his staff was struck by the point of intersection, and here a second stake was fixed in the ice. In this way the process was continued until the line of stakes was completed.

Before quitting the station, a plummet was suspended from a hook directly underneath the centre of the theodolite, and the place where the point touched the ground was distinctly marked. To measure the motion of the line of stakes, we returned to the place a day or two afterwards, and by means of the plummet were able to make the theodolite occupy the exact position which it occupied when the line was set out. The telescope being directed upon the point at the opposite side of the valley, and gradually lowered, it was found that no single stake along the line preserved its first position: they had all shifted downwards. The assistant was sent to the first stake; the point which it had first occupied was again determined, and its present distance from that point accurately measured. The same thing was done in the case of each stake, and thus the displacement of the whole row of stakes was ascertained.* The time at which the stake was fixed, and at which its displacement was measured, being carefully noted, a simple calculation determined *the daily motion* of the stake.

Thus, on the 17th of July, 1857, we set out our first line across the Mer de Glace, at some distance below the Montanvert; on the day following we measured the progress of the stakes. The observed displacements are set down in the following table:—

FIRST LINE.—DAILY MOTION.

No. of stake.		Inches.	No. of stake.		Inches.
WEST 1	moved	12¼	6	moved	..
2	„	16¾	7	„	26¼
3	„	22½	8	„	..
4	„	..	9	„	28¾
5	„	24½	10	„	35½ EAST.

* Great care is necessary on the part of the man who measures the displacements. The staff ought to be placed along the original line, and the assistant ought to walk along it until the foot of a *perpendicular* from the stake is attained. When several days' motion is to be measured, this precaution is absolutely necessary; the eye being liable to be grossly deceived in *guessing* the direction of a perpendicular.

The theodolite in this case stood on the Montanvert side of the valley, and the stakes are numbered from this side. We see that the motion gradually augments from the 1st stake onward—the 1st stake being held back by the friction of the ice against the flanking mountain-side. The stakes 4, 6, and 8 have no motion attached to them, as an accident rendered the measurement of their displacements uncertain. But one remarkable fact is exhibited by this line; the 7th stake stood upon the *middle* of the glacier, and we see that its motion is by no means the quickest; it is exceeded in this respect by the stakes 9 and 10.

The portion of the glacier on which the 10th stake stood was very much cut up by crevasses, and, while the assistant was boring it with his auger, the ice beneath him was observed, through the telescope, to slide suddenly forward for about 4 inches. The other stakes retained their positions, so that the movement was purely local. Deducting the 4 inches thus irregularly obtained, we should have a daily motion of $31\frac{1}{2}$ inches for stake No. 10. The place was watched for some time, but the slipping was not repeated; and a second measurement on the succeeding day made the motion of the 10th stake 32 inches, whilst that of the centre of the glacier was only 27.

Here, then, was a fact which needed explanation; but, before attempting this, I resolved, by repeated measurements in the same locality, to place the existence of the fact beyond doubt. We therefore ascended to a point upon the old and now motionless moraine, a little above the Montanvert Hôtel; and choosing, as before, a well-defined object at the opposite side of the valley, we set between it and the theodolite a row of twenty stakes across the glacier. Their motions, measured on a subsequent day, and reduced to their daily rate, gave the results set down in the following table:—

SECOND LINE.—DAILY MOTION.

No. of stake.		Inches.	No. of stake.		Inches.
WEST 1	moved	7½	11	moved	21
2	,,	10¾	12	,,	22½
3	,,	12¼	13	,,	21
4	,,	14½	14	,,	22½
5	,,	16	15	,,	20½
6	,,	16¾	16	,,	21¾
7	,,	17½	17	,,	22¼
8	,,	19	18	,,	25¼
9	,,	19½	19	,,	
10	,,	21	20	,,	25¾ EAST.

As regards the retardation of the side, we observe here the same fact as that revealed by our first line—the motion gradually augments from the first stake to the last. The stake No. 20 stood upon the dirty portion of the ice, which was derived from the Talèfre tributary of the Mer de Glace, and far beyond the middle of the glacier. These measurements, therefore, corroborate that made lower down, as regards the non-coincidence of the point of swiftest motion with the centre of the glacier.

But it will be observed that the measurements do not show any retardation of the ice at the eastern extremity of the line of stakes—the motion goes on augmenting from the first stake to the last. The reason of this is, that in neither of the cases recorded were we able to get the line quite across the glacier; the crevasses and broken ice-ridges, which intercepted the vision, compelled us to halt before we came sufficiently close to the eastern side to make its retardation sensible. But on the 20th of July my friend Hirst sought out an elevated station on the Chapeau, or eastern side of the valley, whence he could command a view from side to side over all the humps and inequalities of the ice, the fixed point at the opposite side, upon which the telescope was directed, being the corner of a window of the Montanvert Hotel. Along this line were

placed twelve stakes, the daily motions of which were
found to be as follows :—

THIRD LINE.—DAILY MOTION.

No. of stake.		Inches.	No. of stake.		Inches.
EAST 1	moved	19½	7	moved	24½
2	„	22¾	8	„	25
3	„	28¾	9	„	25
4	„	30¼	10	„	18
5	„	33¾	11	„	..
6	„	28¼	12	„	8½ WEST.

The numbering of the stakes along this line commenced
from the Chapeau-side of the glacier, and the retardation
of that side is now manifest enough; the motion gradually
augmenting from 19½ to 33½ inches.　But, comparing the
velocity of the two extreme stakes, we find that the retardation of stake 12 is much greater than that of stake 1.
Stake 5, moreover, which moved with the *maximum* velocity, was not upon the centre of the glacier, but much
nearer to the eastern than to the western side.

It was thus placed beyond doubt that the point of maximum motion of the Mer de Glace, at the place referred to,
is not the centre of the glacier.　But, to make assurance
doubly sure, I examined the comparative motion along
three other lines, and found in all the same undeviating
result.

This result is not only unexpected, but is quite at
variance with the opinions hitherto held regarding the
motion of the Mer de Glace.　The reader knows that the
trunk-stream is composed of three great tributaries from
the Géant, the Léchaud, and the Talèfre.　The Glacier du
Géant fills more than half of the trunk-valley, and the
junction between it and its neighbours is plainly marked
by the dirt upon the surface of the latter.　In fact four
medial moraines are crowded together on the eastern side

of the glacier, and before reaching the Montanvert they have strewn their débris quite over the adjacent ice. A distinct limit is thus formed between the clean Glacier du Géant and the other dirty tributaries of the trunk-stream.

Now the eastern side of the Mer de Glace is observed on the whole to be much more fiercely torn than the western side, and this excessive crevassing has been referred to *the swifter motion of the Glacier du Géant*. It has been thought that, like a powerful river, this glacier drags its more sluggish neighbours after it, and thus tears them in the manner observed. But the measurement of the foregoing three lines shows that this cannot be the true cause of the crevassing. In each case the stakes which moved quickest *lay upon the dirty portion of the trunk-stream*, far to the east of the line of junction of the Glacier du Géant, which in fact moved slowest of all.

The general view of the glacier, and of the shape of the valley which it filled, suggested to me that the analogy with a river might perhaps make itself good beyond the limits hitherto contemplated. The valley was not straight, but sinuous. At the Montanvert the convex side of the glacier was turned eastward; at some distance higher up, near the passages called *Les Ponts*, it was turned westward; and higher up again it was turned once more, for a long stretch, eastward. Thus between Trélaporte and the Ponts we had what is called a point of contrary flexure, and between the Ponts and the Montanvert a second point of the same kind.

Supposing a river, instead of the glacier, to sweep through this valley; *its* point of maximum motion would not always remain central, but would deviate towards that side of the valley to which the river turned its convex boundary. Indeed the positions of towns along the banks of a navigable river are mainly determined by this circum-

stance. They are, in most cases, situate on the convex
sides of the bends, where the rush of the water prevents
silting up. Can it be then that the ice exhibits a similar deportment? that the same principle which regulates
the distribution of people along the banks of the Thames
is also acting with silent energy amid the glaciers of the
Alps? If this be the case, the position of the point of
maximum motion ought, of course, to shift with the
bending of the glacier. Opposite the Ponts, for example,
the point ought to be on the Glacier du Géant, and westward of the centre of the trunk-stream ; while, higher up,
we ought to have another change to the eastern side, in
accordance with the change of flexure.

On the 25th of July a line was set out across the glacier,
one of its fixed termini being a mark upon the first of the
three Ponts. The motion of this line, measured on a sub-
quent day, and reduced to its daily rate, was found to
be as follows :—

FOURTH LINE.—DAILY MOTION.

No. of stake.		Inches.	No. of stake.		Inches.
EAST 1	moved	$6\frac{1}{4}$	10	moved	21
2	,,	8	11	,,	$20\frac{1}{2}$
3	,,	$12\frac{1}{2}$	12	,,	$23\frac{1}{4}$
4	,,	$15\frac{1}{4}$	13	,,	$23\frac{1}{4}$
5	,,	$15\frac{1}{2}$	14	,,	21
6	,,	$18\frac{3}{4}$	15	,,	$22\frac{1}{4}$
7	,,	$18\frac{1}{4}$	16	,,	$17\frac{1}{4}$
8	,,	$18\frac{3}{4}$	17	,,	15 WEST.
9	,,	$19\frac{1}{2}$			

This line, like the third, was set out and numbered from
the eastern side of the glacier, the theodolite occupying a
position on the heights of the Echelets. A moment's
inspection of the table reveals a fact different from that
observed on the third line ; *there* the most easterly stake
moved with more than twice the velocity of the most

westerly one; *here*, on the contrary, the most westerly stake moves with more than twice the velocity of the most easterly one.

To enable me to compare the motion of the eastern and western halves of the glacier with greater strictness, my able and laborious companion undertook the task of measuring with a surveyor's chain the line just referred to; noting the pickets which had been fixed along the line, and the other remarkable objects which it intersected. The difficulty of thus directing a chain over crevasses and ridges can hardly be appreciated except by those who have tried it. Nevertheless, the task was accomplished, and the width of the Mer de Glace, at this portion of its course, was found to be 863 yards, or almost exactly half a mile.

Referring to the last table, it will be seen that the two stakes numbered 12 and 13 moved with a common velocity of 23¼ inches per day, and that their motion is swifter than that of any of the others. The point of swiftest motion may be taken midway between them, and this point was found by measurement to lie 233 yards *west* of the dirt which marked the junction of the Glacier du Géant with its fellow tributaries: whereas, in the former cases, it lay a considerable distance *east* of this limit. Its distance from the eastern side of the glacier was 601 yards, and from the western side 262 yards, being 170 yards west of the centre of the glacier.

But the measurements enabled me to take the stakes in pairs, and to compare the velocity of a number of them which stood at certain distances from the eastern side of the valley, with an equal number which stood *at the same distances* from the western side. By thus arranging the points two by two, I was able to compare the motion of the entire body of the ice at the one side of the central line with that of the ice at the other side. Stake 17 stood about as far from the western side of the glacier as stake 3 did from its eastern side; 16 occupied the same relation to 4; 15, to 5; 13, to 7; and 12, to 9.

Calling each pair of points which thus stand at equal distances from the opposite sides *corresponding points*, the following little table exhibits their comparative motions:—

NUMBERS AND VELOCITIES OF CORRESPONDING POINTS ON THE FOURTH LINE.

	No.	Vel.	No.	Vel.	No.	Vel.	No.	Vel.	No.	Vel.
West..	17	15	16	17¼	15	22[illegible]	13	23¾	12	23¼
East ..	3	12½	4	15¼	5	15½	7	18¼	9	19½

The table explains itself. We see that while stake 17, which stands *west* of the centre, moves 15 inches, stake 3, which stands an equal distance *east* of the centre, moves only 12½ inches. Comparing every pair of the other points, we find the same to hold good; the western stake moves in each case faster than the corresponding eastern one. Hence, *the entire western half of the Mer de Glace, at the place crossed by our fourth line, moves more quickly than the eastern half of the glacier.*

We next proceeded farther up, and tested the contrary curvature of the glacier, opposite to Trélaporte. The station chosen for this purpose was on a grassy platform of the promontory, whence, on the 28th of July, a row of stakes was fixed at right angles to the axis of the glacier. Their motions, measured on the 31st, gave the following results:—

FIFTH LINE.*—DAILY MOTION.

No. of stake.		Inches.	No. of stake.		Inches.
WEST. 1	moved	11¼	9	moved	19¾
2	,,	13½	10	,,	19
3	,,	12¼	11	,,	19½
4	,,	15	12	,,	17½
5	,,	15¼	13	,,	16
6	,,	16	14	,,	14¾
7	,,	17¼	15	,,	10 EAST.
8	,,	19¼			

* The details of the measurement of the fourth and fifth lines are published in the 'Philosophical Transactions,' vol. cxlix. p. 261.

This line was set out and numbered from the Trélaporte side of the valley, and was also measured by Mr. Hirst, over boulders, ice-ridges, chasms, and moraines. The entire width of the glacier here was found to be 893 yards, or somewhat wider than it is at the Ponts. It will also be observed that its motion is somewhat slower.

An inspection of the notes of this line showed me that stakes 3 and 14, 4 and 12, 7 and 10, were "corresponding points;" the first of each pair standing as far from the western side, as the second stood from the eastern. In the following table these points and their velocities are arranged exactly as in the case of the fourth line.

NUMBERS AND VELOCITIES OF THE CORRESPONDING POINTS ON THE FIFTH LINE.

	No.	Vel.	No.	Vel.	No.	Vel.
West ..	3	12¾	4	15	7	17¼
East ..	14	14¾	12	17½	10	19

In each case we find that the stake on the eastern side moves more quickly than the corresponding one upon the western side: so that where the fifth line crosses the glacier *the eastern half of the Mer de Glace moves more quickly than the western half.* This is the reverse of the result obtained at our fourth line, but it agrees with that obtained on our first three lines, where the curvature of the valley is similar. The analogy between a river and a glacier moving through a sinuous valley is therefore complete.

Supposing the points of maximum motion to be determined for a great number of lines across the glacier, the line uniting all these points is what mathematicians would call the *locus* of the point of maximum motion. At Trélaporte this line would lie east of the centre; at the Ponts it would lie west of the centre; hence, in passing from Trélaporte to the Ponts, it must cross the axis of the glacier. Again, at the Montanvert, it would lie east of the centre, and between the Ponts and the Montanvert the axis of the glacier would be crossed a second time. Supposing the dotted line in Fig. 21 to represent the middle

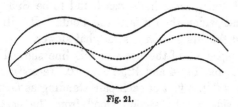

Fig. 21.

line of the glacier, then the defined line would represent the locus of the point of maximum motion. *It is a curve more deeply sinuous than the valley itself, and it crosses the axis of the glacier at each point of contrary flexure.*

To complete our knowledge of the motion of the Mer de Glace, we afterwards determined the velocity of its two accessible tributaries — the Glacier du Géant, and the Glacier de Léchaud. On the 29th of July, a line of stakes was set out across the former, a little above the Tacul, and their motion was subsequently found to be as follows:

SIXTH LINE. —DAILY MOTION.

No. of stake.		Inches.	No. of stake.		Inches.
1	moved	11	6	moved	12¾
2	„	10	7	„	10½
3	„	12	8	„	10
4	„	13	9	„	9
5	„	12	10	„	5

The width of the glacier at this place we found to be 1134 yards, and its maximum velocity, as shown by the foregoing table, 13 inches a day.

On the 1st of August a line was set out across the Glacier de Léchaud, above its junction with the Talèfre: it commenced beneath the block of stone known as the Pièrre de Béranger. The displacements of the stakes, measured upon the 3rd of August, gave the following results :—

SEVENTH LINE.—DAILY MOTION.

No. of stake.		Inches.	No. of stake.		Inches.
1	moved	4½	6	moved	7½
2	„	8¼	7	„	6¼
3	„	9½	8	„	8½
4	„	9	9	„	7
5	„	8½	10	„	5½

The width of the Glacier de Léchaud at this place was found to be 825 yards; its maximum motion, as shown by the table, being 9½ inches a-day. This is the slowest rate which we observed upon either the Mer de Glace or its tributaries. The width of the Talèfre-branch, as it descends the cascade, or, in other words, before it is influenced by the pressure of the Léchaud, was found approximately to be 638 yards.

The widths of the tributaries were determined for the purpose of ascertaining the amount of lateral compression endured by the ice in its passage through the neck of the valley at Trélaporte. Adding all together we have—

Géant	1134 yards.
Léchaud	825 „
Talèfre	638 „
Total	2597 yards.

These three branches, as shown by the actual measurement of our 5th line, are forced at Trélaporte through a channel 893 yards wide; the width of the trunk stream is a little better than one-third of that of its tributaries, and it passes through this gorge at a velocity of nearly 20 inches a-day.

Limiting our view to one of the tributaries only, the result is still more impressive. Previous to its junction with the Talèfre, the Glacier de Léchaud stretches before the observer as a broad river of ice, measuring 825 yards across: at Trélaporte it is squeezed, in a frozen vice, between the Talèfre on one side and the Géant on the other, to a driblet, measuring 85 yards in width, or about one-tenth of its former transverse dimension. It will of course be understood that it is the *form* and not the *volume* of the glacier that is affected to this enormous extent by the pressure.

Supposing no waste took place, the Glacier de Léchaud would force precisely the same amount of ice through the "narrows" at Trélaporte, in one day, as it sends past the Pìerre de Béranger. At the latter place its velocity is about half of what it is at the former, but its width is more than nine times as great. Hence, if no waste took place, its *depth*, at Trélaporte, would be at *least* 4½ times its depth opposite the Pìerre de Béranger. Superficial and sub-glacial melting greatly modify this result. Still I think it extremely probable that observations directed to this end would prove the comparative shallowness of the upper portions of the Glacier de Léchaud.

ICE-WALL AT THE TACUL.

VELOCITIES OF TOP AND BOTTOM.

(11.)

As regards the motion of the *surface* of a glacier, two laws
are to be borne in mind: 1st, that regarding the quicker
movement of the centre; 2nd, that regarding the locus
of the point of maximum motion.   Our next care must
be to compare the motion of the surface of a glacier
with the motion of those parts which lie near its bed.
Rendu first surmised that the bottom of the glacier was
retarded by friction, and both Professor Forbes * and M.
Martins† have confirmed the conjecture.   Theirs are the
only observations which we possess upon the subject; and
I was particularly desirous to instruct myself upon this
important head by measurements of my own.

During the summer of 1857 the eastern side of the
Glacier du Géant, near the Tacul, exposed a nearly vertical precipice of ice, measuring 140 feet from top to
bottom.   I requested Mr. Hirst to fix two stakes in the
same vertical plane, one at the top of the precipice
and one near the bottom.   This he did upon the 3rd of
August, and on the 5th I accompanied him to measure
the progress of the stakes.   On the summit of the precipice, and running along it, was the lateral moraine of
the glacier.   The day was warm and the ice liquefying
rapidly, so that the boulders and débris, deprived incessantly of their support, came in frequent leaps and rushes
down the precipice.   Into this peril my guide was about

* 'Edinb. Phil. Journ.,' Oct. 1846, p. 417.
† Agassiz, 'Système Glaciaire,' p. 522.

to enter, to measure the displacement of the lower stake, while I was to watch, and call out the direction in which he was to run when a stone gave way. But I soon found that the initial motion was no sure index of the final motion. By striking the precipice, the stones were often deflected, and carried wide of their original direction. I therefore stopped the man, and sent him to the summit of the precipice to remove all the more dangerous blocks. This accomplished, he descended, and, while I stood beside him, executed the required measurement. From the 3rd to the 5th of August the upper stake had moved twelve inches, and the lower one six.

Unfortunately some uncertainty attached itself to this result, due to the difficulty of fixing the lower stake. The guide's attention had been divided between his work and his safety, and he had to retreat more than a dozen times from the falling boulders and débris. I, on the other hand, was unwilling to accept an observation of such importance with a shade of doubt attached to it. Hence arose the desire to measure the motion myself. On the 11th of August I therefore reascended to the Tacul, and fixed a stake at the top of the precipice, and another at the bottom. While sitting on the old moraine looking at the two pickets, the importance of determining the motion of a point midway between the top and bottom forcibly occurred to me, but, on mentioning it to my guide, he promptly pronounced any attempt of the kind absurd.

On scanning the place carefully, however, the value of the observation appeared to me to outweigh the amount of danger. I therefore took my axe, placed a stake and an auger against my breast, buttoned my coat upon them, and cut an oblique staircase up the wall of ice, until I reached a height of forty feet from the bottom. Here the position of the stake being determined by Mr. Hirst, who was at the theodolite, I pierced the ice with the auger,

drove in the stake, and descended without injury. During
the whole operation however my guide growled audibly.

On the following morning we commenced the ascent of
Mont Blanc, a narrative of which is given in Part I.
We calculated on an absence of three days, and estimated
that the stakes which had just been fixed would be ready
for measurement on our return; but we did not reach
Chamouni until the afternoon of Friday, the 14th. Heavy
clouds settled, during our descent, upon the summits behind us, and a thunder-peal from the Aiguilles soon
heralded a fall of rain, which continued without intermission till the afternoon of the 16th, when the atmosphere cleared, and showed the mountains clothed to their
girdles with snow. The Montanvert was thickly covered,
and on our way to it we met the servants in charge of
the cattle, which had been driven below the snow-line to
obtain food.

On Monday morning, the 17th, a dense fog filled the
valley of the Mer de Glace. I watched it anxiously. The
stakes which we had set at the Tacul had been often
in my thoughts, and I wished to make some effort to
save the labour and peril incurred in setting them
from being lost. I therefore set out, in one of the clear
intervals, accompanied by my friend and Simond, determined to measure the motion of the stakes, if possible,
or to fix them more firmly, if they still stood. As we
passed, however, from l'Angle to the glacier, the fog
became so dense and blinding that we halted. At my request Mr. Hirst returned to the Montanvert; and Simond,
leaving the theodolite in the shelter of a rock, accompanied me through the obscurity to the Tacul. We found
the topmost stake still stuck by its point in the ice; but
the two others had disappeared, and we afterwards discovered their fragments in a snow-buttress, which reared
itself against the base of the precipice. They had been

o 2

hit by the falling stones, and crushed to pieces. Having thus learned the worst, we descended to the Montanvert amid drenching rain.

On the morning of the 18th there was no cloud to be seen anywhere, and the sunlight glistened brightly on the surface of the ice. We ascended to the Tacul. The spontaneous falling of the stones appeared more frequent this morning than I had ever seen it. The sun shone with unmitigated power upon the ice, producing copious liquefaction. The rustle of falling débris was incessant, and at frequent intervals the boulders leaped down the precipice, and rattled with startling energy amid the rocks at its base. I sent Simond to the top to remove the looser stones; he soon appeared, and urged the moraine-shingle in showers down the precipice, upon a bevelled slope of which some blocks long continued to rest. They were out of the reach of the guide's bâton, and he sought to dislodge them by sending other stones down upon them. Some of them soon gave way, drawing a train of smaller shingle after them; others required to be hit many times before they yielded, and others refused to be dislodged at all. I then cut my way up the precipice in the manner already described, fixed the stake, and descended as speedily as possible. We afterwards fixed the bottom stake, and on the 20th the displacements of all three were measured.* The spaces passed over by the respective stakes in 24 hours were found to be as follows:—

	Inches.
Top stake	6·00
Middle stake	4·50
Bottom stake	2·56

The height of the precipice was 140·8 feet, but it sloped off at its upper portion. The height of the middle

* On this latter occasion my guide volunteered to cut the steps for me up to the pickets; and I permitted him to do so. In fact, he was at last as anxious as myself to see the measurement carried out.

stake above the ground was 35 feet, and of the bottom
one 4 feet. It is therefore proved by these measurements
that the bottom of the ice-wall at the Tacul moves with
less than half the velocity of the top; while the displacement of the intermediate stake shews how the velocity
gradually increases from the bottom upwards.

WINTER MOTION OF THE MER DE GLACE.

(12.)

THE winter measurements were executed in the manner
already described, on the 28th and 29th of December,
1859.   The theodolite was placed on the mountain's
side flanking the glacier, and a well-defined object was
chosen at the opposite side of the valley, so that a
straight line between this object and the theodolite was
approximately perpendicular to the axis of the glacier.
Fixing the telescope in the first instance with its cross hairs
upon the object, its end was lowered until it struck the
point upon the glacier at which a stake was to be fixed.
Thanks to the intelligence of my assistants, after the fixing
of the first stake they speedily took up the line at all
other points, requiring very little correction to make their
positions perfectly accurate.   On the day following that
on which the stakes were driven in, the theodolite was
placed in the same position, and the distances to which the
stakes had moved from their original positions were accurately determined.   As already stated, the first line crossed
the glacier about 80 yards above the Montanvert Hotel.

LINE NO. I.—WINTER MOTION IN TWENTY-FOUR HOURS.

No. of stake.		Inches.	No. of stake.		Inches.
WEST 1	..	$7\frac{1}{4}$	7	..	$15\frac{3}{4}$
2	..	11	8	..	$15\frac{3}{4}$
3	..	$13\frac{1}{2}$	9	..	$12\frac{1}{4}$
4	..	13	10	..	12
5	..	$13\frac{3}{4}$	11	..	$6\frac{1}{2}$ EAST.
6	..	$14\frac{1}{4}$			

The maximum here is fifteen and three-quarters inches;
the maximum summer motion of the same portion of the

glacier is about thirty inches. These measurements also
show that in winter, as well as in summer, the side of the
glacier opposite to the Montanvert moves quicker than
that adjacent to it. The stake which moved with the maximum velocity was beyond the moraine of La Noire. The
second line crossed the glacier about 130 yards below the
Montanvert.

LINE No. II.—WINTER MOTION IN TWENTY-FOUR HOURS.

No. of stake.	Inches.	No. of stake.	Inches.
1 ..	7¾	6 ..	15¾
2 ..	9½	7 ..	17½
3 ..	13¾	8 ..	16½
4 ..	16	9 ..	14½
5 ..	16	10 ..	14

The maximum here is an inch and three-quarters greater
than that of line No. 1. The summer maximum at this
portion of the glacier also exceeds that of the part intersected by line No. 1. The surface of the glacier between
both lines is in a state of tension which relieves itself by a
system of transverse fissures, and thus permits of the
quicker advance of the forward portion.

My desire, in making these measurements, was, in the
first place, to raise the winter observations of the motion to
the same degree of accuracy as that already enjoyed by
the summer ones. Auguste Balmat had already made a
series of winter observations on the Mer de Glace; but
they were made in the way employed before the introduction of the theodolite by Agassiz and Forbes, and
shared the unavoidable roughness of such a mode of measurement. They moreover gave us no information as to
the motion of the different parts of the glacier along the
same transverse line, and this, for reasons which will
appear subsequently, was the point of chief interest to me.

CAUSE OF GLACIER-MOTION.

DE SAUSSURE'S THEORY.

(13.)

PERHAPS the first attempt at forming a glacier-theory is
that of Scheuchzer in 1705.   He supposed the motion to
be caused by the conversion of water into ice within the
glacier; the known and almost irresistible expansion which
takes place on freezing, furnishing the force which pushed
the glacier downward.   This idea was illustrated and developed with so much skill by M. de Charpentier, that his
name has been associated with it; and it is commonly
known as the Theory of Charpentier, or the Dilatation-
Theory.   M. Agassiz supported this theory for a time, but
his own thermometric experiments show us that the body
of the glacier is at a temperature of 32° Fahr.; that consequently there is no interior magazine of cold to freeze the
water with which the glacier is supposed to be incessantly
saturated.   So that these experiments alone, if no other
grounds existed, would prove the insufficiency of the theory
of dilatation.   I may however add, that the arguments most
frequently urged against this theory deal with an assumption, which I do not think its author ever intended to
make.

Another early surmise was that of Altmann and Grüner
(1760), both of whom conjectured that the glacier slid
along its bed.   This theory received distinct expression
from De Saussure in 1799 ; and has since been associated
with the name of that great alpine traveller, being usually
called the 'Theory of Saussure,' and sometimes the 'Sliding Theory.'   It is briefly stated in these words :—

"Almost every glacier reposes upon an inclined bed, and those of any considerable size have beneath them, even in winter, currents of water which flow between the ice and the bed which supports it. It may therefore be understood that these frozen masses, drawn down the slope on which they repose, disengaged by the water from all adhesion to the bottom, sometimes even raised by this water, must glide by little and little, and descend, following tho inclinations of the valleys, or of the slopes which they cover. It is this slow but continual sliding of the ice on its inclined base which carries it into the lower valleys."*

De Saussure devoted but little time to the subject of glacier-motion; and the absence of completeness in the statement of his views, arising no doubt from this cause, has given subsequent writers occasion to affix what I cannot help thinking a strained interpretation to the sliding theory. It is alleged that he regarded a glacier as a perfectly rigid body; that he considered it to be "a mass of ice of small depth, and considerable but uniform breadth, sliding down a uniform valley, or pouring from a narrow valley into a wider one."† The introduction "of the smallest flexibility or plasticity" is moreover emphatically denied to him. ‡

It is by no means probable that the great author of the 'Voyages' would have subscribed to this "rigid" annotation. His theory, be it remembered, is to some extent *true:* the glacier moves over its bed in the manner supposed, and the rocks of Britain bear to this day the traces of these mighty sliders. De Saussure probably contented himself with a general statement of what he believed to

* 'Voyages,' § 535.
† Forbes's ' Occ. Pap., p. 100.
‡ "I adhere to the definition as excluding the introduction of the smallest flexibility or plasticity." 'Occ. Pap.,' p. 96.

be the substantial cause of the motion. He visited the Jardin, and saw the tributaries of the Mer de Glace turning round corners, welding themselves together, and afterwards moving through a sinuous trunk-valley; and it is scarcely credible that in the presence of such facts he would have denied all flexibility to the glacier.

The statement that he regarded a glacier to be a mass of ice of uniform width, is moreover plainly inconsistent with the following description of the glacier of Mont Dolent: —" Its most elevated plateau is a great circus, surrounded by high cliffs of granite, of pyramidal forms; thence the glacier descends through a gorge, in which *it is narrowed;* but after having passed the gorge, it *enlarges again,* spreading out like a fan. Thus it has on the whole the form of a sheaf tied in the middle and dilated at its two extremities." *

Curiously enough this very glacier, and these very words, are selected by M. Rendu as illustrative of the plasticity of glaciers. "Nothing," he says, "shows better the extent to which a glacier moulds itself to its locality than the form of the glacier of Mont Dolent in the Valley of Ferret;" and he adds, in connexion with the same passage, these remarkable words :—" There is a multitude of facts which would seem to necessitate the belief that the substance of glaciers enjoys a kind of ductility which permits it to mould itself to the locality which it occupies, to grow thin, to swell, and to narrow itself like a soft paste." †

* ' Voyages,' tome ii. p. 290.

† In connexion with this brief sketch of the ' Sliding Theory,' it ought to be stated, that Mr. Hopkins has proved experimentally, that ice may descend an incline at a sensibly uniform rate, and that the velocity is augmented by increasing the weight. In this remarkable experiment the motion was due to the slow disintegration of the lower surface of the ice. See ' Phil. Mag.,' 1845, vol. 26.

RENDU'S THEORY.

(14.)

M. Rendu, Bishop of Annecy, to whose writings I have just referred, died last autumn. He was a man of great repute in his diocese, and we owe to him one of the most remarkable essays upon glaciers that have ever appeared. His knowledge was extensive, his reasoning close and accurate, and his faculty of observation extraordinary. With these were associated that intuitive power, that presentiment concerning things as yet untouched by experiment, which belong only to the higher class of minds. Throughout his essay a constant effort after quantitative accuracy reveals itself. He collects observations, makes experiments, and tries to obtain numerical results; always taking care, however, so to state his premises and qualify his conclusions that nobody shall be led to ascribe to his numbers a greater accuracy than they merit. It is impossible to read his work, and not feel that he was a man of essentially truthful mind, and that science missed an ornament when he was appropriated by the Church.

The essay above referred to is printed in the tenth volume of the Memoirs of the Royal Academy of Sciences of Savoy, published in 1841, and is entitled, ' *Théorie des Glaciers de la Savoie, par M. le Chanoine Rendu, Chevalier du Mérite Civil et Secrétaire perpétuel.*' The paper had been written for nearly two years, and might have remained unprinted, had not another publication on the same subject called it forth.

I will place a few of the leading points of this remarkable production before the reader; commencing with a generalization which is highly suggestive of the character of the author's mind.

He reflects on the accumulation of the mountain-snows, each year adding fifty-eight inches of ice to a glacier. This would make Mont Blanc four hundred feet higher in a century, and four thousand feet higher in a thousand years. "It is evident," he says, "that nothing like this occurs in nature." The escape of the ice then leads him to make some general remarks on what he calls the "law of circulation." "The conserving will of the Creator has employed for the permanence of His work the great law of *circulation*, which, strictly examined, is found to reproduce itself in all parts of nature. The waters circulate from the ocean to the air, from the air to the earth, and from the earth to the ocean. . . . The elements of organic substances circulate, passing from the solid to the liquid or aëriform condition, and thence again to the state of solidity or of organisation. That universal agent which we designate by the names of fire, light, electricity, and magnetism, has probably also a *circulation* as wide as the universe." The italics here are Rendu's own. This was published in 1841, but written, we are informed, nearly two years before. In 1842 Mr. Grove wrote thus:—" Light, heat, magnetism, motion, and chemical affinity, are all convertible material affections." More recently Helmholtz, speaking of the "circuit" formed by "heat, light, electricity, magnetism, and chemical affinity," writes thus:—"Starting from each of these different manifestations of natural forces, we can set every other in action." I quote these passages because they refer to the same agents as those named by M. Rendu, and to which he ascribes "*circulation*." Can it be doubted that this Savoyard priest had a premonition of the Conservation of Force? I do not want to lay more stress than it deserves upon a conjecture of this kind; but its harmony with an essay remarkable for its originality gives it a significance which, if isolated, it might not possess.

With regard to the glaciers, Rendu commences by dividing them into two kinds, or rather the self-same glacier into two parts, one of which he calls the "*glacier réservoir,*" the other the "*glacier d'écoulement,*"—two terms highly suggestive of the physical relationship of the *névé* and the glacier proper. He feeds the reservoirs from three sources, the principal one of which is the snow, to which he adds the rain, and the vapours which are condensed upon the heights without passing into the state of either rain or snow. The conversion of the snow into ice he supposes to be effected by four different causes, the most efficacious of which is *pressure.** It is needless to remark that this quite agrees with the views now generally entertained.

In page 60 of the volume referred to there is a passage which shows that the "veined structure" of the glacier had not escaped him, though it would seem that he ascribed it to stratification. "When," he writes, "we perceive the profile of a glacier on the walls of a crevasse, we see different layers distinct in colour, but more particularly in density; some seem to have the hardness, as they have the greenish colour, of glass; others preserve the whiteness and porosity of the snow." There is also a very close resemblance between his views of the influence of "time and cohesion" and those of Prof. Forbes. "We may conclude," he writes, "that *time,* favouring the action of *affinity,* and the pressure of the layers one upon the other, causes the little crystals of which snow is composed to approach each other, bring them into contact, and convert them into ice." † Regelation also appears to have attracted his notice. ‡ "When we fill an ice-house," he writes, "we break the ice into very small fragments; afterwards we wet it with water 8 or 10 degrees above zero (Cent.) in temperature; but, notwithstanding this, the whole

* 'Memoir,' p. 77.   † P. 75.   ‡ P. 71.

is converted into a compact mass of ice." He moreover maintains, in almost the same language as Prof. Forbes,[*] the opinion, that ice has always an inner temperature lower than zero (Cent.). He believed this to be a property "inherent to ice." "Never," he says, "can a calorific ray pass the first surface of ice to raise the temperature of the interior."[†]

He notices the direction of the glacier as influencing the wasting of its ridges by the sun's heat; ascribing to it the effect to which I have referred in explaining the wave-like forms upon the surface of the Mer de Glace. His explanation of the Moulins, too, though insufficient, assigns a true cause, and is an excellent specimen of physical reasoning.

With regard to the diminution of the *glaciers réservoirs*, or, in other words, to the manner in which the ice disappears, notwithstanding the continual additions made to it, we have the following remarkable passage:—"In seeking the cause of the diminution of glaciers, it has occurred to my mind that the ice, notwithstanding its hardness and its rigidity, can only support a given pressure without breaking or being squeezed out. According to this supposition, whenever the pressure exceeds that force, there will be rupture of the ice, and a flow in consequence. Let us take, at the summit of Mont Blanc, a column of ice reposing on a horizontal base. The ice which forms the first layer of that column is compressed by the weight of all the layers above it; but if the solidity of the said first layer can only support a weight equal to 100, when the weight exceeds this amount there will be rupture and spreading out of the ice of the base. Now, something very similar occurs in the immense crust of ice which covers the summits of Mont Blanc. This crust appears to augment at the upper surface and to diminish by the sides. To assure

oneself that the movement is due to the force of pressure,
it would be necessary to make a series of experiments upon
the solidity of ice, such as have not yet been attempted." *
I may remark that such experiments substantially verify
M. Rendu's notion.

But it is his observations and reasoning upon the *glaciers
d'écoulement* that chiefly interest us. The passages in
his writings where he insists upon the power of the
glaciers to mould themselves to their localities, and compares them to a soft paste, to lava at once ductile and
liquid, are well known from the frequent and flattering,
references of Professor Forbes; but there are others of
much greater importance, which have hitherto remained
unknown in this country. Regarding the motion of the
Mer de Glace, Rendu writes as follows :—

"I sought to appreciate the quantity of its motion; but
I could only collect rather vague data. I questioned my
guides regarding the position of an enormous rock at the
edge of the glacier, but still upon the ice, and consequently partaking of its motion. The guides showed
me the place where it stood the preceding year, and
where it had stood two, three, four, and five years previously; they showed me the place where it would be
found in a year, in two years, &c. ; *so certain are they
of the regularity of the motion.* Their reports, however,
did not always agree precisely with each other, and their
indications of time and distance lack the precision without
which we proceed obscurely in the physical sciences. In
reducing these different indications to a mean, I found
the total advance of the glacier to be about 40 feet
a year. During my last journey I obtained more certain
data, which I have stated in the preceding chapter. *The
enormous difference between both results arises from the fact
that the latter observations were made at the centre of the*

* Page 80.

glacier, WHICH MOVES MORE RAPIDLY, *while the former were made at the side, where the ice* IS RETAINED BY THE FRICTION AGAINST ITS ROCKY WALLS." *

An opinion, founded on a grave misapprehension which Rendu enables us to correct, is now prevalent in this country, not only among the general public, but also among those of the first rank in science. The nature of the mistake will be immediately apparent. At page 128 of the 'Travels in the Alps' its distinguished author gives a sketch of the state of our knowledge of glacier-motion previous to the commencement of his inquiries. He cites Ebel, Hugi, Agassiz, Bakewell, De la Beche, Shirwell, Rendu, and places them in open contradiction to each other. Rendu, he says, gives the motion of the Mer de Glace to be "242 feet per annum; 442 feet per annum; a a foot a day; 400 feet per annum, and 40 feet per annum, or *one-tenth* of the last!".... and he adds, "I was not therefore wrong in supposing that the actual progress of a glacier was yet a new problem when I commenced my observations on the Mer de Glace in 1842." †

In the 'North British Review' for August, 1859, a writer equally celebrated for the brilliancy of his discoveries and the vigour of his pen, collected the data furnished by the above paragraph into a table, which he introduced to his readers in the following words:—"It is to Professor Forbes alone that we owe the first and most correct researches respecting the motion of glaciers; and in proof of this, we have only to give the following list of observations which had been previously made.

* Page 95.
† At page 38 of the 'Travels' the following passage also occurs:—"I believe that I may safely affirm that not one observation of the rate of motion of a glacier, either on the average or at any particular season of the year, existed when I commenced my experiments in 1842."

Observers.	Name of glacier.	Annual rate of motion.
Ebel	Chamouni	14 feet
Ebel	Grindelwald	25 „
Hugi	Aar	240 „
Agassiz	Aar	200 „
Bakewell	Mer de Glace	540 „
De la Beche	Mer de Glace	600 „
Shirwell	Mer de Glace	300 „
M. Rendu	Mer de Glace	365 „
Saussure's Ladder ..	Mer de Glace	375 „

... Such was the state of our knowledge when Professor Forbes undertook the investigation of the subject."

I am persuaded that the writer of this article will be the first to applaud any attempt to remove an error which, advanced on his great authority, must necessarily be widely disseminated. The numbers in the above table certainly differ widely, and it is perhaps natural to conclude that such discordant results can be of no value; but the fact really is that *every one of them may be perfectly correct*. This fact, though overlooked by Professor Forbes, was clearly seen by Rendu, who pointed out with perfect distinctness the sources from which the discrepancies were derived.

"It is easy," he says, "to comprehend that it is impossible to obtain a general measure,—that there ought to be one for each particular glacier. The nature of the slope, the number of changes to which it is subjected, the depth of the ice, the width of the couloir, the form of its sides, and a thousand other circumstances, must produce variations in the velocity of the glacier, and these circumstances cannot be everywhere absolutely the same. Much more, it is not easy to obtain this velocity for a single glacier, and for this reason. In those portions where the inclination is steep, the layer of ice is thin, and its velocity is great; in those where the slope is almost nothing, *the glacier swells and accumulates;* the mass in motion being double, triple, &c., the motion is only the half, the third, &c.

"But this is not all," adds M. Rendu: "*Between the Mer de Glace and a river, there is a resemblance so complete that it is impossible to find in the latter a circumstance which does not exist in the former.* In currents of water the motion is not uniform, neither throughout their width nor throughout their depth; *the friction of the bottom, that of the sides,* the action of obstacles, cause the motion to vary, *and only towards the middle of the surface is this entire . . . .*" *

In 1845 Professor Forbes appears to have come to the same conclusion as M. Rendu; for after it had been proved that the centre of the Aar glacier moved quicker than the side in the ratio of fourteen to one, he accepted the result in these words:—"The movement of the centre of the glacier is to that of a point five mètres from the edge as FOURTEEN to ONE: such is the effect of plasticity!" †   Indeed, if the differences exhibited in the table were a proof of error, the observations of Professor Forbes himself would fare very ill.   The measurements of glacier-motion made with his own hands vary from less than 42 feet a year to 848 feet a year, the minimum being less than *one-twentieth* of the maximum; and if we include the observations made by Balmat, the fidelity of which has been certified by Professor Forbes, the minimum is only *one-thirty-seventh* of the maximum.

There is another point connected with Rendu's theory which needs clearing up:—"The idea," writes the eminent reviewer, "that a glacier is a semifluid body is no doubt startling, especially to those who have seen the apparently rigid ice of which it is composed.   M. Rendu himself shrank from the idea, and did not scruple to say that 'the rigidity of a mass of ice was in direct opposition to it;' and we think that Professor Forbes himself must have stood aghast when his fancy first associated the notion of imperfect fluidity with the solid or even the fissured ice of

* 'Théorie,' p. 96.   † 'Occ. Pap.,' p. 74.

the glacier, and when he saw in his mind's eye the glaciers of the Alps flowing like a river along their rugged bed. A truth like this was above the comprehension and beyond the sympathy of the age; and it required a moral power of no common intensity to submit it to the ordeal of a shallow philosophy, and the sneers of a presumptuous criticism."

These are strong words; but the fact is that, so far from " shrinking" from the idea, Rendu affirmed, with a clearness and an emphasis which have not been exceeded since, that all the phenomena of a river were reproduced upon the Mer de Glace; its deeps, its shallows, its widenings, its narrowings, its rapids, its places of slow motion, and the quicker flow of its centre than of its sides. He did not shrink from accepting a difference between the central and lateral motion amounting to a ratio of ten to one—a ratio so large that Professor Forbes at one time regarded the acceptance of it as a simple absurdity. In this he was perhaps justified; for his own first observations, which, however valuable, were hasty and incomplete, gave him a maximum ratio of about one and a half to one, while the ratio in some cases was nearly one of *equality*. The observations of Agassiz however show that the ratio, instead of being ten to one, may be *infinity* to one; for the lateral ice may be so held back by a local obstacle that in the course of a year it shall make no sensible advance at all.

From one thing only did M. Rendu shrink; and it is *the* thing regarding which we are still disunited. He shrank from stating the physical quality of the *ice* in virtue of which a *glacier* moved like a river. He demands experiments upon snow and ice to elucidate this subject. The very observations which Professor Forbes regards as proofs are those of which we require the physical explanation. It is not the viscous flow, if you please to call it such, of the glacier as a whole that here concerns us; but it is the quality of the *ice* in virtue of which this kind of motion is accomplished. Professor Forbes sees this difference clearly enough: he speaks of "fissured ice" being "flexible" in hand specimens; he compares the glacier to a mixture of ice and sand; and finally, in a more matured paper, falls back for an explanation upon the observations of Agassiz regarding the capillaries of the glacier.*

———————

(15.)

The measurements of Agassiz and Forbes completely verify the anticipations of Rendu; but no writer with whom I am acquainted has added anything essential to the Bishop's statements as to the identity of glacier and liquid motion. He laid down the conditions of the problem with perfect clearness, and, as regards the distribution of merit, the point to be decided is the relative importance of his idea, and of the measurements which were subsequently made.

The observations on which Professor Forbes based the analogy between a glacier and a river are the following:— In 1842 he fixed four marks upon the Mer de Glace a little below the Montanvert, the first of which was 100 yards distant from the side of the glacier, while the last was at

* In all that has been written upon glaciers in this country the above passages from the writings of Rendu are unquoted; and many who mingled very warmly in the discussions of the subject were, until quite recently, ignorant of their existence. I was long in this condition myself, for I never supposed that passages which bear so directly upon a point so much discussed, and of such cardinal import, could have been overlooked; or that the task of calling attention to them should devolve upon myself nearly twenty years after their publication. Now that they are discovered, I conceive no difference of opinion can exist as to the propriety of placing them in their true position.

the centre "or a little beyond it." The relative velocity
of these four points was found to be

1·000   1·332   1·356   1·367.

The first observations were made upon two of these points,
two others being subsequently added. Professor Forbes
also determined the velocity of two points on the Glacier
du Géant, and found the ratio of motion, in the first
instance, to be as 14 to 32. Subsequent measurements,
however, showed the ratio to be as 14 to 18, the larger
motion belonging to the station nearest to the centre of
the glacier. These are the only measurements which I
can find in his large work that establish the swifter motion
of the centre of the glacier; and in these cases the velocity
of the centre is compared with that of *one side* only. In
no instance that I am aware of, either in 1842 or subsequent years, did Professor Forbes extend his measurements quite across a glacier; and as regards completeness
in this respect, no observations hitherto made can at all
compare with those executed at the instance of Agassiz
upon the glacier of the Aar.

In 1844 Professor Forbes made a series of interesting
experiments on a portion of the Mer de Glace near
l'Angle. He divided a length of 90 feet into 45 equal
spaces, and fixed pins at the end of each. His theodolite
was placed upon the ice, and in seventeen days he found
that the ice 90 feet nearer the centre than the theodolite
had moved 26 inches past the latter. These measurements
were undertaken for a special object, and completely
answered the end for which they were intended.

In 1846 Professor Forbes made another important
observation. Fixing three stakes at the heights of 8, 54,
and 143 feet above the bed of the glacier, he found that
in five days they moved respectively 2·87, 4·18, and
4·66 feet. The stake nearest the bed moved most slowly,

thus showing that the ice is retarded by friction. This result was subsequently verified by the measurements of M. Martins, and by my own.

If we add to the above an observation made during a short visit to the Aletsch glacier in 1844, which showed its lateral retardation, I believe we have before us the whole of the measurements executed by Professor Forbes, which show the analogy between the motion of a glacier and that of a viscous body.

Illustrative of the same point, we have the elaborate and extensive series of measurements executed by M. Wild under the direction of M. Agassiz upon the glacier of the Aar in 1842, 1843, 1844, and 1845, which exhibit on a grand scale, and in the most conclusive manner, the character of the motion of this glacier; and also show, on close examination, an analogy with fluid motion which neither M. Agassiz nor Professor Forbes suspected. The former philosopher publishes a section in his 'Système Glaciaire,' entitled 'Migrations of the Centre;' in which he shows that the middle of the glacier is not always the point of swiftest motion. The detection of this fact demonstrates the attention devoted by M. Agassiz to the discussion of his observations, but he gives no clue to the cause of the variation. On inspecting the shape of the valley through which the Aar glacier moves, I find that these "migrations" follow the law established in 1857 upon the Mer de Glace, and enunciated at page 286.

To sum up this part of the question :—The *idea* of semifluid motion belongs entirely to Rendu; the *proof* of the quicker central flow belongs in part to Rendu, but almost wholly to Agassiz and Forbes; the proof of the retardation of the bed belongs to Forbes alone; while the discovery of the locus of the point of maximum motion belongs, I suppose, to me.

FORBES'S THEORY.

(16.)

THE formal statement of this theory is given in the following words:—"A glacier is an imperfect fluid, or viscous
body, which is urged down slopes of a certain inclination
by the mutual pressure of its parts." The consistency of
the glacier is illustrated by reference to treacle, honey,
and tar, and the theory thus enunciated and exemplified is
called the ' Viscous Theory.'

It has been the subject of much discussion, and great
differences of opinion are still entertained regarding it.
Able and sincere men take opposite sides; and the extraordinary number of Reviews which have appeared upon the
subject during the last two years show the interest which
the intellectual public of England take in the question.
The chief differences of opinion turn upon the inquiry as
to what Professor Forbes really meant when he propounded
the viscous theory; some affirm one thing, some another,
and, singularly enough, these differences continue, though
the author of the theory has at various times published
expositions of his views.

The differences referred to arise from the circumstances
that a sufficient distinction has not been observed between
facts and *principles*, and that the viscous theory has assumed
various forms since its first promulgation. It has been
stated to me that the theory of Professor Forbes is "the
congeries of facts" which he has discovered. But it is
quite evident that no recognition, however ample, of these
facts would be altogether satisfactory to Professor Forbes
himself. He claims recognition of his *theory*,* and no writer

* "Mr. Hopkins," writes Professor Forbes, "has done me the honour,
in the memoirs before alluded to, to mention with approbation my observa

with whom I am acquainted makes such frequent use of
the term. What then can the viscous theory mean apart
from the facts? I interpret it as furnishing the principle
from which the facts follow as physical consequences—that
the *glacier* moves as a river because the *ice* is viscous. In
this sense only can Professor Forbes's views be called a
theory; in any other, his experiments are mere illustrations
of the facts of glacier motion, which do not carry us a hair's
breadth towards their physical cause.

What then is the meaning of viscosity or viscidity? I
have heard it defined by men of high culture as "gluey
tenacity;" and such tenacity they once supposed a glacier to
possess. If we dip a spoon into treacle, honey, or tar, we can
draw the substance out into filaments, and the same may be
done with melted caoutchouc or lava. All these substances
are viscous, and all of them have been chosen to illustrate
the physical property in virtue of which a glacier moves.
Viscosity then consists in the power of being drawn out
when subjected to a force of tension, the substance, after
stretching, being in a state of molecular equilibrium, or, in
other words, devoid of that elasticity which would restore it
to its original form. This certainly was the idea attached
to Professor Forbes's words by some of his most strenuous
supporters, and also by eminent men who have never
taken part in any controversy on the subject. Mr. Darwin,
for example, speaks of felspathic rocks being "stretched"
while flowing slowly onwards in a pasty condition, in precisely the same manner as Professor Forbes believes that
the ice of moving glaciers is stretched and fissured; and
Professor Forbes himself quotes these words of Mr. Darwin
as illustrative of his theory.*

tions and experiments on the subject of glaciers. He has been more
sparing either in praise or criticism of the theory which I have founded
upon them. Had Mr. Hopkins," &c.—*Eighth Letter;* 'Occ. Papers,' p. 66.
    * 'Occ. Papers,' p. 92.

The question now before us is,—Does a glacier exhibit that power of yielding to a force of tension which would entitle its ice to be regarded as a viscous substance?

With a view to the solution of this question Mr. Hirst took for me the inclinations of the Mer de Glace and all its tributaries in 1857; the effect of a change of inclination being always noted. I will select from those measurements a few which bear more specially upon the subject now under consideration, commencing with the Glacier des Bois, down which the ice moves in that state of wild dislocation already described. The inclination of the glacier above this cascade is 5° 10′, and that of the cascade itself is 22° 20′, the change of inclination being therefore 17° 10′.

In Fig. 22 I have protracted the inclination of the cascade and of the glacier above it; the line A B representing the

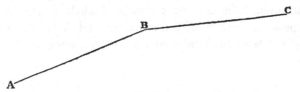

Fig. 22.

former and B C the latter. Now a stream of molten lava, of treacle, or tar, would, in virtue of its viscosity, be able to flow over the brow at B without breaking across; but this is not the case with the glacier; it is so smashed and riven in crossing this brow, that, to use the words of Professor Forbes himself, "it pours into the valley beneath in a cascade of icy fragments."

But this reasoning will appear much stronger when we revert to other slopes upon the Mer de Glace. For example, its inclination above l'Angle is 4°, and it afterwards descends a slope of 9° 25′, the change of inclination being 5° 25′. If we protract these inclinations to scale, we have

P

the line A B, Fig. 23, representing the steeper slope, and B C that of the glacier above it.   One would surely think

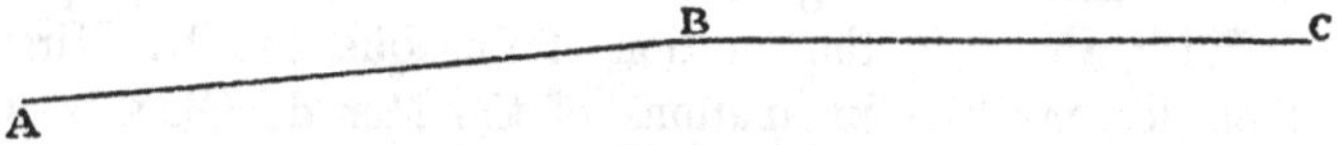

Fig. 23.

that a viscous body could cross the brow B without trans-verse fracture, but this the glacier cannot do, and Professor Forbes himself pronounces this portion of the Mer de Glace impassable.   Indeed it was the profound crevasses here formed which placed me in a difficulty already referred to.   Higher up again, the glacier is broken on passing from a slope of 3° 10′ to one of 5°.   Such observations show how differently constituted a glacier is from a stream of lava in a "pasty condition," or of treacle, honey, tar, or melted caoutchouc, to all which it has been compared.   In the next section I shall endeavour to explain the origin of the crevasses, and shall afterwards make a few additional remarks on the alleged viscosity of ice.

THE CREVASSES.

(17.)

HAVING made ourselves acquainted with the motion of the glacier, we are prepared to examine those rents, fissures, chasms, or, as they are most usually called, *Crevasses*, by which all glaciers are more or less intersected. They result from the motion of the glacier, and the laws of their formation are deduced immediately from those of the motion. The crevasses are sometimes very deep and numerous, and apparently without law or order in their distribution. They cut the ice into long ridges, and break these ridges transversely into prisms; these prisms gradually waste away, assuming, according to the accidents of their melting, the most fantastic forms. I have seen them like the mutilated statuary of an ancient temple, like the crescent moon, like huge birds with outstretched wings, like the claws of lobsters, and like antlered deer.   Such fantastic sculpture is often to be found on the ice cascades, where the riven glacier has piled vast blocks on vaster pedestals, and presented them to the wasting action of sun and air.   In Fig. 24 I have given a sketch of a mass of ice of this character, which stood in 1859 on the dislocated slope of the Glacier des Bois.

It is usual for visitors to the Montanvert to descend to the glacier, and to be led by their guides to the edges of the crevasses, where, being firmly held, they look down into them; but those who have only made their acquaintance in this way know but little of their magnitude and beauty in the more disturbed portions of glaciers. As might be expected, they have been the graves of many a mountaineer; and the skeletons found upon the glacier prove that even the chamois itself, with its elastic muscles and

admirable sureness of foot, is not always safe among the
crevasses. They are grandest in the higher ice-regions,

Fig. 24.

where the snow hangs like a coping over their edges, and
the water trickling from these into the gloom forms splendid
icicles. The Görner Glacier, as we ascend it towards the
old Weissthor, presents many fine examples of such crevasses; the ice being often torn in a most curious and
irregular manner. You enter a porch, pillared by icicles,
and look into a cavern in the very body of the glacier,
encumbered with vast frozen bosses which are fringed all
round by dependent icicles. At the peril of your life from
slipping, or from the yielding of the stalactites, you may
enter these caverns, and find yourself steeped in the blue
illumination of the place. Their beauty is beyond description; but you cannot deliver yourself up, heart and soul, to
its enjoyment. There is a strangeness about the place which
repels you, and not without anxiety do you look from your

ledge into the darkness below, through which the sound of
subglacial water sometimes rises like the tolling of distant bells. You feel that, however the cold splendours
of the place might suit a purely spiritual essence, they
are not congenial to flesh and blood, and you gladly
escape from its magnificence to the sunshine of the world
above.

From their numbers it might be inferred that the
formation of crevasses is a thing of frequent occurrence
and easy to observe; but in reality it is very rarely observed. Simond was a man of considerable experience upon
the ice, but the first crevasse he ever saw formed was during
the setting out of one of our lines, when a narrow rent
opened beneath his feet, and propagated itself through the
ice with loud cracking for a distance of 50 or 60 yards. Crevasses always commence in this way as mere narrow cracks,
which open very slowly afterwards. I will here describe
the only case of crevasse-forming which has come under
my direct observation.

On the 31st of July, 1857, Mr. Hirst and myself, having
completed our day's work, were standing together upon
the Glacier du Géant, when a loud dull sound, like that
produced by a heavy blow, seemed to issue from the body
of the ice underneath the spot on which we stood. This
was succeeded by a series of sharp reports, which were
heard sometimes above us, sometimes below us, sometimes
apparently close under our feet, the intervals between the
louder reports being filled by a low singing noise. We
turned hither and thither as the direction of the sounds
varied; for the glacier was evidently breaking beneath
our feet, though we could discern no trace of rupture.
For an hour the sounds continued without our being able
to discover their source; this at length revealed itself
by a rush of air-bubbles from one of the little pools upon
the surface of the glacier, which was intersected by

the newly-formed crevasse. We then traced it for some
distance up and down, but hardly at any place was it
sufficiently wide to permit the blade of my penknife
to enter it. M. Agassiz has given an animated description of the terror of his guides upon a similar occasion,
and there was an element of awe in our own feelings
as we heard the evening stillness of the glacier thus
disturbed.

With regard to the mechanical origin of the crevasses
the most vague and untenable notions had been entertained
until Mr. Hopkins published his extremely valuable
papers. To him, indeed, we are almost wholly indebted
for our present knowledge of the subject, my own experiments upon this portion of the glacier-question being for
the most part illustrations of the truth of his reasoning.
To understand the fissures in their more complex aspects
it is necessary that we should commence with their
elements. I shall deal with the question in my own
way, adhering, however, to the mechanical principles upon
which Mr. Hopkins has based his exposition.

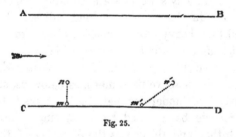

Fig. 25.

Let A B, C D, be the bounding sides of a glacier moving
in the direction of the arrow ; let m, n be two points upon
the ice, one, m, close to the retarding side of the valley,
and the other, n, at some distance from it. After a
certain time, the point m will have moved downwards to
m', but in consequence of the swifter movement of the

parts at a distance from the sides, *n* will have moved in
the same time to *n'*. Thus the line *m n*, instead of being
at right angles to the glacier, takes up the oblique position
m' n'; but to reach from *m'* to *n'* the line *m n* would
have to stretch itself considerably; every other line that
we can draw upon the ice parallel to *m' n'* is in a similar
state of tension; or, in other words, the sides of the glacier
are acted upon by an oblique pull towards the centre.
Now, Mr. Hopkins has shown that the direction in which
this oblique pull is strongest encloses an angle of 45° with
the side of the glacier.

What is the consequence of this? Let A B, C D, fig. 26,

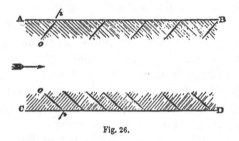

Fig. 26.

represent, as before, the sides of the glacier, moving
in the direction of the arrow; let the shading lines
enclose an angle of 45° with the sides. *Along* these lines
the marginal ice suffers the greatest strain, and, consequently *across* these lines and at right angles to them, the
ice tends to break and to form *marginal crevasses*. The
lines, *o p, o p*, mark the direction of these crevasses; they
are at right angles to the line of greatest strain, and
hence also enclose an angle of 45° with the side of the
valley, *being obliquely pointed upwards*.

This latter result is noteworthy; it follows from the
mechanical data that the swifter motion of the centre
tends to produce marginal crevasses which are inclined
from the side of the glacier towards its source, and not

towards its lower extremity. But when we look down upon a glacier thus crevassed, the first impression is that the sides have been *dragged down*, and have left the central portions behind them; indeed, it was this very appearance that led M. de Charpentier and M. Agassiz into the error of supposing that the sides of a glacier moved more quickly than its middle portions; and it was also the delusive aspect of the crevasses which led Professor Forbes to infer the slower motion of the eastern side of the Mer de Glace.

The retardation of the ice is most evident near the sides; in most cases, the ice for a considerable distance right and left of the central line moves with a sensibly uniform velocity; there is no dragging of the particles asunder by a difference of motion, and, consequently, a compact centre is perfectly compatible with fissured sides. Nothing is more common than to see a glacier with its sides deeply cut, and its central portions compact; this, indeed, is always the case where the glacier moves down a bed of uniform inclination.

But supposing that the bed is not uniform—that the valley through which the glacier moves changes its inclination abruptly, so as to compel the ice to pass over a brow; the glacier is then circumstanced like a stick which we try to break by holding its two ends and pressing it against the knee. The brow, where the bed changes its inclination, represents the knee in the case of the stick, while the weight of the glacier itself is the force that tends to break it. It breaks; and fissures are formed across the glacier, which are hence called *transverse crevasses*.

No glacier with which I am acquainted illustrates the mechanical laws just developed more clearly and fully than the Lower glacier of Grindelwald. Proceeding along the ordinary track beside the glacier, at about an hour's distance from the village the traveller reaches a point whence a view of the glacier is obtained from the heights above it. The marginal fissures are very cleanly cut, and point nearly in the direction already indicated; the glacier also changes its inclination several times along the distance within the observer's view. On crossing each brow the glacier is broken across, and a series of transverse crevasses is formed, which follow each other down the slope. At the bottom of the slope tension gives place to pressure, the walls of the crevasses are squeezed together, and the chasms closed up. They remain closed along the comparatively level space which stretches between the base of one slope and the brow of the next; but here the glacier is again transversely broken, and continues so until the base of the second slope is reached, where longitudinal pressure instead of longitudinal strain begins to act, and the fissures are closed as before. In Fig. 27A I have given a sketchy section of a portion of the glacier, illustrating the formation of the crevasses at the top of a slope, and their subsequent obliteration at its base.

Another effect is here beautifully shown, namely, the union of the transverse and marginal crevasses to form continuous fissures which stretch quite across the glacier. Fig. 27B will illustrate my meaning, though very imperfectly; it represents a plan of a portion of the Lower Grindelwald glacier, with both marginal and transverse fissures drawn upon it. I have placed it under the section so that each part of it may show in plan the portion of the glacier which is shown in section immediately above it. It shows how the marginal crevasses remain after the compression of the centre has obliterated the transverse ones; and how the latter join on to the former, so as to form continuous fissures, which sweep across the glacier in vast curves, with their convexities turned upwards.

The illusion before referred to is here strengthened; the crevasses turn, so to say, *against* the direction of motion,

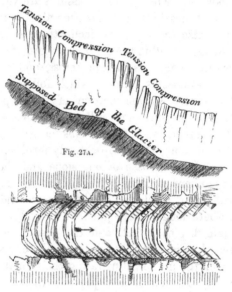

Fig. 27A.

Fig. 27B.

instead of forming loops, with their convexities pointing downwards, and thus would impress a person unacquainted with the mechanical data with the idea that the glacier margins moved more quickly than the centre. The figures are intended to convey the idea merely; on the actual slopes of the glacier between twenty and thirty chasms may be counted: the word "compression" also ought to have been limited to the level portions of the sketch.

Besides the two classes of fissures mentioned we often find others, which are neither marginal nor transverse. The terminal portions of many glaciers, for example, are in a state of compression; the snout of the glacier abuts against the ground, and having to bear the thrust of the mass behind it, if it have room to expand laterally, the ice

will yield, and *longitudinal crevasses* will be formed. They
are of very common occurrence, but the finest example
of the kind is perhaps exhibited by the glacier of the
Rhone. After escaping from the steep gorge which holds
the cascade, this glacier encounters the bottom of a comparatively wide and level valley; the resistance to its forward motion is augmented, while its ability to expand
laterally is increased; it has to bear a longitudinal thrust,
and it splits at right angles to the pressure. A series of
fissures is thus formed, the central ones of which are truly
longitudinal; but on each side of the central line the
crevasses diverge, and exhibit a fan-like arrangement. This
disposition of the fissures is beautifully seen from the summit of the Mayenwand on the Grimsel Pass.

Here then we have the elements, so to speak, of glacier-
crevassing, and through their separate or combined action
the most fantastic cutting up of a glacier may be effected.
And see how beautifully these simple principles enable
us to account for the remarkable crevassing of the
eastern side of the Mer
de Glace. Let A B, C D,
be the opposite sides of a
portion of the glacier, near
the Montanvert; C D being
east, and A B west, the
glacier moving in the di-
rection of the arrow; let

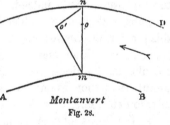

Fig. 28.

the points *m n* represent the extremities of our line of
stakes, and let us suppose an elastic string stretched across
the glacier from one to the other. We have proved that
the point of maximum motion here lies much nearer to
the side C D than to A B. Let *o* be this point, and, seizing
the string at *o*, let it be drawn in the direction of motion
until it assumes the position, *m, o′, n*. It is quite evident
that *o′ n* is in a state of greater tension than *o′ m*, and the

ice at the eastern side of the Mer de Glace is in a precisely
similar mechanical condition. It suffers a greater strain
than the ice at the opposite side of the valley, and hence is
more fissured and broken. Thus we see that the crevassing
of the eastern side of the glacier is a simple consequence
of the quicker motion of that side, and does not, as hitherto
supposed, demonstrate its slower motion. The reason why
the eastern side of the glacier, as a whole, is much more
fissured than the western side is, that there are two long
segments which turn their convex curvature eastward, and
only one segment of the glacier which turns its convexity
westward.

The lower portion of the Rhone glacier sweeps round
the side of the valley next the Furca, and turns throughout a convex curve to this side: the crevasses here are
wide and frequent, while they are almost totally absent at
the opposite side of the glacier. The lower Grindelwald
glacier turns at one place a convex curve towards the
Eiger, and is much more fissured at that side than at
the opposite one; indeed, the fantastic ice-splinters,
columns, and minarets, which are so finely exhibited upon
this glacier, are mainly due to the deep crevassing of the
convex side. Numerous other illustrations of the law might,
I doubt not, be discovered, and it would be a pleasant and
useful occupation to one who takes an interest in the subject, to determine, by strict measurements upon other glaciers, the locus of the point of maximum motion, and
to observe the associated mechanical effects.

The appearance of crevasses is often determined by circumstances more local and limited than those above indicated; a boss of rock, a protuberance on the side of the
flanking mountain, anything, in short, which checks the
motion of one part of the ice and permits an adjacent
portion to be pushed away from it, produces crevasses.
Some valleys are terminated by a kind of mountain-circus

with steep sides, against which the snow rises to a considerable height. As the mass is urged downwards, the lower portion of the snow-slope is often torn away from its higher portion, and a chasm is formed, which usually extends round the head of the valley. To such a crevasse the specific name *Bergschrund* is applied in the Bernese Alps; I have referred to one of them in the account of the "Passage of the Strahleck."

(18.)

The phenomena described and accounted for in the last chapter have a direct bearing upon the question of viscosity. In virtue of the quicker central flow the lateral ice is subject to an oblique strain; but, instead of stretching, it breaks, and marginal crevasses are formed. We also see that a slight curvature in the valley, by throwing an additional strain upon one half of the glacier, produces an augmented crevassing of that side.

But it is known that a substance confessedly viscous may be broken by a sudden shock or strain. Professor Forbes justly observes that sealing-wax at moderate temperatures will mould itself (with time) to the most delicate inequalities of the surface on which it rests, but may at the same time be shivered to atoms by the blow of a hammer. Hence, in order to estimate the weight of the objection that a glacier breaks when subjected to strain, we must know the conditions under which the force is applied.

The Mer de Glace has been shown (p. 287) to move through the neck of the valley at Trélaporte at the rate of twenty inches a day. Let the sides of this page represent the boundaries of the glacier at Trélaporte, and any one of its lines of print a transverse slice of ice. Supposing the line to move down the page as the slice of ice moves down the valley, then the bending of the ice in twenty-four hours, shown on such a scale, would only be sufficient to push forward the centre in advance of the sides by a very small fraction of the width of the line of print. To such an extremely gradual strain the ice is unable to accommodate itself without fracture.

Or, referring to actual numbers:—the stake No. 15 on our 5th line, page 284, stood on the lateral moraine of the Mer de Glace; and between it and No. 14 a distance of 190 feet intervened. Let A B, Fig. 29, be the side of the glacier, moving in the direction of the arrow, and let $a\ b\ c\ d$ be a square upon the glacier with a side of 190 feet. The whole square moves with the ice, but the side $b\ d$ moves quickest; the point a moving 10 inches, while b moves 14·75 inches in 24 hours; the differential motion therefore amounts to an inch in five hours. Let $ab'\ d'c$ be the shape of the figure after five hours' motion; then the line $a\ b$ would be extended to $a\ b'$ and cd to cd'.

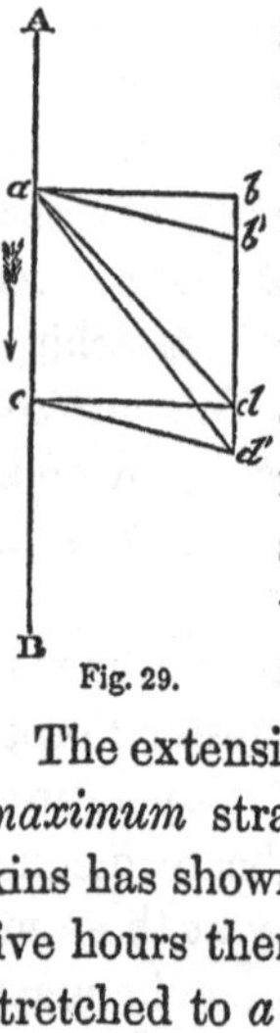

Fig. 29.

The extension of *these* lines does not however express the *maximum* strain to which the ice is subjected. Mr. Hopkins has shown that this takes place along the line ad; in five hours then this line, if capable of stretching, would be stretched to ad'. From the data given every boy who has mastered the 47th Proposition of the First Book of Euclid can find the length both of ad and ad'; the former is 3224·4 inches, and the latter is 3225·1, the difference between them being seven-tenths of an inch.

This is the amount of yielding required from the ice in five hours, but it cannot grant this; the glacier breaks, and numerous marginal crevasses are formed. It must not be forgotten that the evidence here adduced merely shows what ice cannot do; what it *can* do in the way of viscous yielding we do not know: there exists as yet no single experiment on great masses or small to show that ice possesses in any sensible degree that power of being drawn out which seems the very essence of viscosity.

I have already stated that the crevasses, on their first formation, are exceedingly narrow rents, which widen very slowly. The new crevasse observed by our guide required several days to attain a width of three inches; while that observed by Mr. Hirst and myself did not widen a single inch in three days. This, I believe, is the general character of the crevasses; they form suddenly and open slowly. Both facts are at variance with the idea that ice is viscous; for were this substance capable of stretching at the slow rate at which the fissures widen, there would be no necessity for their formation.

It cannot be too clearly and emphatically stated that the *proved* fact of a glacier conforming to the law of semi-fluid motion is a thing totally different from the *alleged* fact of its being viscous. Nobody since its first enunciation disputed the former. I had no doubt of it when I repaired to the glaciers in 1856; and none of the eminent men who have discussed this question with Professor Forbes have thrown any doubt upon his measurements. It is the assertion that small pieces of ice are proved to be viscous * by the experiments made upon glaciers, and the consequent impression left upon the public mind—that ice possesses the "gluey tenacity" which the term viscous suggests—to which these observations are meant to apply.

* "The viscosity, though it cannot be traced in the parts *if very minute*, nevertheless *exists* there, as unequivocally proved by experiments on the large scale."—Forbes in 'Phil. Mag.,' vol. x., p. 301.

HEAT AND WORK.

(19.)

GREAT scientific principles, though usually announced by individuals, are often merely the distinct expression of thoughts and convictions which had long been entertained by all advanced investigators. Thus the more profound philosophic thinkers had long suspected a certain equivalence and connexion between the various forces of nature ; experiment had shown the direct connexion and mutual convertibility of many of them, and the spiritual insight, which, in the case of the true experimenter, always surrounds and often precedes the work of his hands, revealed more or less plainly that natural forces either had a common root, or that they formed a circle, whose links were so connected that by starting from any one of them we could go through the circuit, and arrive at the point from which we set out.   For the last eighteen years this subject has occupied the attention of some of the ablest natural philosophers, both in this country and on the Continent.   The connexion, however, which has most occupied their minds is that between *heat* and *work;* the absolute numerical equivalence of the two having, I believe, been first announced by a German physician named Meyer, and experimentally proved in this country by Mr. Joule.

A lead bullet may be made hot enough to burn the hand, by striking it with a hammer, or by rubbing it against a board; a clever blacksmith can make a nail red-hot by hammering it; Count Rumford boiled water by the heat developed in the boring of cannon, and inferred from the experiment that heat was not what it was generally

supposed to be, an imponderable fluid, but a kind of
motion generated by the friction. Now Mr. Joule's
experiments enable us to state the exact amount of heat
which a definite expenditure of mechanical force can originate. I say *originate*, not drag from any hiding-place in
which it had concealed itself, but actually bring into existence, so that the total amount of heat in the universe is
thereby augmented. If a mass of iron fall from a tower
770 feet in height, we can state the precise amount of heat
developed by its collision with the earth. Supposing all
the heat thus generated to be concentrated in the iron
itself, its temperature would thereby be raised nearly 10°
Fahr. Gravity in this case has expended a certain amount
of force in pulling the iron to the earth, and this force is
the *mechanical equivalent* of the heat generated. Further-
more, if we had a machine so perfect as to enable us to
apply all the heat thus produced to the raising of a weight,
we should be able, by it, to lift the mass of iron to the
precise point from which it fell.

But the heat cannot lift the weight and still continue
heat; this is the peculiarity of the modern view of the
matter. The heat is consumed, used up, it is no longer
heat; but instead of it we have a certain amount of gravi-
tating force stored up, which is ready to act again, and to
regenerate the heat when the weight is let loose. In fact,
when the falling weight is stopped by the earth, the motion
of its mass is converted into a motion of its molecules;
when the weight is lifted by heat, molecular motion is converted into ordinary mechanical motion, but for every
portion of either of them brought into existence an equivalent portion of the other must be consumed.

What is true for masses is also true for atoms. As the
earth and the piece of iron mutually attract each other,
and produce heat by their collision, so the carbon of a
burning candle and the oxygen of the surrounding air

mutually attract each other; they rush together, and on collision the arrested motion becomes heat. In the former case we have the conversion of gravity into heat, in the latter the conversion of chemical affinity into heat; but in each case the process consists in the generation of motion by attraction, and the subsequent change of that motion into motion of another kind. Mechanically considered, the attraction of the atoms and its results is precisely the same as the attraction of the earth and weight and *its* results.

But what is true for an atom is also true for a planet or a sun. Supposing our earth to be brought to rest in her orbit by a sudden shock, we are able to state the exact amount of heat which would be thereby generated. The consequence of the earth's being thus brought to rest would be that it would fall into the sun, and the amount of heat which would be generated by this second collision is also calculable. Helmholtz has calculated that in the former case the heat generated would be equal to that produced by the combustion of fourteen earths of solid coal, and in the latter case the amount would be 400 times greater.

Whenever a weight is lifted by a steam-engine in opposition to the force of gravity an amount of heat is consumed equivalent to the work done; and whenever the molecules of a body are shifted in opposition to their mutual attractions work is also performed, and an equivalent amount of heat is consumed. Indeed the amount of work done in the shifting of the molecules of a body by heat, when expressed in ordinary mechanical work, is perfectly enormous. The lifting of a heavy weight to the height of 1000 feet may be as nothing compared with the shifting of the atoms of a body by an amount so small that our finest means of measurement hardly enable us to determine it. Different bodies give heat different degrees of trouble, if I may use the term, in shifting their atoms and putting them in new places.

Iron gives more trouble than lead; and water gives far more trouble than either. The heat expended in this molecular work is lost as heat; it does not show itself as temperature. Suppose the heat produced by the combustion of an ounce of candle to be concentrated in a pound of iron, a certain portion of that heat would go to perform the molecular work to which I have referred, and the remainder would be expended in raising the temperature of the body; and if the same amount of heat were communicated to a pound of iron and to a pound of lead, the balance in favour of temperature would be greater in the latter case than in the former, because the heat would have less molecular work to do; the lead would become more heated than the iron. To raise a pound of iron a certain number of degrees in temperature would, in fact, require more than three times the absolute quantity of heat which would be required to raise a pound of lead the same number of degrees. Conversely, if we place the pound of iron and the pound of lead, heated to the same temperature, into ice, we shall find that the quantity of ice melted by the iron will be more than three times that melted by the lead. In fact, the greater amount of molecular work invested in the iron now comes into play, the atoms again obey their own powerful forces, and an amount of heat corresponding to the energy of these forces is generated.

This molecular work is that which has usually been called *specific heat*, or *capacity for heat*. According to the *materialistic* view of heat, bodies are figured as sponges, and heat a kind of fluid absorbed by them, different bodies possessing different powers of absorption. According to the *dynamic* view, as already explained, heat is regarded as a motion, and capacity for heat indicates the quantity of that motion consumed in internal changes.

The greatest of these changes occurs when a body passes from one state of aggregation to another, from the solid to the liquid, or from the liquid to the aëriform state; and the quantity of heat required for such changes is often enormous. To convert a pound of ice at 32° Fahr. into water *at the same temperature* would require an amount of heat competent, if applied as mechanical force, to lift the same pound of ice to a height of 110,000 feet; it would raise a ton of ice nearly 50 feet, or it would lift between 49 and 50 tons to a height of one foot above the earth's surface. To convert a pound of water at 212° into a pound of steam at the same temperature would require an amount of heat which would perform nearly seven times the amount of mechanical work just mentioned.

This heat is entirely expended in *interior work*,* and does nothing towards augmenting the temperature; the water is at the temperature of the ice which produced it, both are 32°; and the steam is at the temperature of the water which produced it, both are 212°. The whole of the heat is consumed in producing the change of aggregation; I say "*consumed*," not hidden or "latent" in either the water or the steam, but absolutely non-existent as heat. The molecular forces, however, which the heat has sacrificed itself to overcome are able to reproduce it; the water in freezing and the steam in condensing give out the exact amount of heat which they consumed when the change of aggregation was in the opposite direction.

At a temperature of several degrees below its freezing point ice is much harder than at 32°. I have more than once cooled a sphere of the substance in a bath of solid carbonic acid and ether to a temperature of 100° below the freezing point. During the time of cooling the ice crackled audibly from its contraction, and afterwards it quite resisted the edge of a knife; while at 32° it may be cut or crushed with extreme facility. The cold sphere was subjected to

---

* I borrow this term from Professor Clausius's excellent papers on the Dynamical Theory of Heat.

pressure; it broke with the detonation of a vitreous body, and was taken from the press a white opaque powder; which, on being subsequently raised to 32° and again compressed, was converted into a pellucid slab of ice.

But before the temperature of 32° is quite attained, ice gives evidence of a loosening of its crystalline texture. Indeed the unsoundness of ice at and near its melting point has been long known. Sir John Leslie, for example, states that ice at 32° is *friable;* and every skater knows how rotten ice becomes before it thaws. M. Person has further shown that the latent heat of ice, that is to say, the quantity of heat necessary for its liquefaction, is not quite expressed by the quantity consumed in reducing ice at 32° to the liquid state. The heat begins to be rendered latent, or in other words the change of aggregation commences, a little before the substance reaches 32°,—a conclusion which is illustrated and confirmed by the deportment of melting ice under pressure.

In reference to the above result Professor Forbes writes as follows:—"I have now to refer to a fact . . . . established by a French experimenter, M. Person, who appears not to have had even remotely in his mind the theory of glaciers, when he announced the following facts, viz.—'That ice does not pass abruptly from the solid to the fluid state; that it begins to *soften* at a temperature of 2° Centigrade below its thawing point; that, consequently, between 28° 4′ and 32° of Fahr. ice is actually passing through various degrees of plasticity within narrower limits, but in the same manner that wax, for example, softens before it melts.'" The "*softening*" here referred to is the "friability," of Sir J. Leslie, and what I have called a "loosening of the texture." Let us suppose the Serpentine covered by a sheet of pitch so smooth and hard as to enable a skater to glide over it; and which is afterwards gradually warmed until it begins to bend under his weight, and finally lets him through.   A comparison of this deportment with that of a sheet of ice under the same circumstances enables us to decide whether ice "passes through various degrees of plasticity in the same manner as wax softens before it melts."   M. Person concerned himself solely with the heat absorbed, and no doubt in both wax and ice that heat is expended in "interior work."   In the one case, however, the body is so constituted that the absorbed heat is expended in rendering the substance viscous; and the question simply is, whether the heat absorbed by the ice gives its molecules a freedom of play which would entitle it also to be called viscous; whether, in short, "rotten ice" and softened wax present the same physical qualities?

(20.)

There is one other point in connexion with the viscous theory which claims our attention.   The announcement of that theory startled scientific men, and for two or three years after its first publication it formed the subject of keen discussion.   This finally subsided, and afterwards Professor Forbes drew up an elaborate paper, which was presented in three parts to the Royal Society in 1845 and 1846, and subsequently published in the 'Philosophical Transactions.'

In the concluding portion of Part III. Professor Forbes states and answers the question, "How far a glacier is to be regarded as a plastic mass?" in these words:—"Were a glacier composed of a solid crystalline cake of ice, fitted or moulded to the mountain bed which it occupies, like a lake tranquilly frozen, it would seem impossible to admit such a flexibility or yielding of parts as should permit any comparison to a fluid or semifluid body, transmitting pressure horizontally, and whose parts might change their mutual positions so that one part should be pushed out whilst another remained behind. But we know, in point of fact, that a glacier is a body very differently constituted. It is clearly proved by the experiments of Agassiz and others that the glacier is not a mass of ice, but of ice and water, the latter percolating freely through the crevices of the former to all depths of the glacier; and it is a matter of ocular demonstration that these crevices, though very minute, communicate freely with one another to great distances; the water with which they are filled communicates force also to great distances, and exercises a tremendous hydrostatic pressure to move onwards in the direction in which gravity urges it, the vast porous mass of seemingly rigid ice in which it is as it were bound up."

"Now the water in the crevices," continues Professor Forbes, "does not constitute the glacier, but only the principal vehicle of the force which acts on it, and the slow irresistible energy with which the icy mass moves onwards from hour to hour with a continuous march, bespeaks of itself the presence of a fluid pressure. But if the ice were not in some degree ductile or plastic, this pressure could never produce any the least forward motion of the mass. The pressure in the capillaries of the glacier can only tend to separate one particle from another, and thus produce tensions and compressions *within the body of the glacier itself*, which yields, owing to its slightly ductile nature, in the direction of least resistance, retaining its continuity, or recovering it by reattachment after its parts have suffered a bruise, according to the violence of the action to which it has been exposed."

I will not pretend to say that I fully understand this passage, but, taking it and the former one together, I think it is clear that the water which is supposed to gorge the capillaries of the glacier is assumed to be essential to its motion. Indeed, an extreme degree of sensitiveness has been ascribed to the glacier as regards the changes of temperature by which the capillaries are affected. In three succeeding days, for example, Professor Forbes found the diurnal summer motion of a point upon the Mer de Glace to increase from 15·2 to 17·5 inches a day; a result which he says he is "persuaded" to be due to the increasing heat of the weather at the time. If, then, the glacier capillaries can be gorged so quickly as this experiment would indicate, it is fair to assume that they are emptied with corresponding speed when the supply is cut away.

The extraordinary coldness of the weather previous to the Christmas of 1859 is in the recollection of everybody: this lowness of temperature also extended to the Mer de Glace and its environs. I had last summer left with Auguste Balmat and the Abbé Vueillet thermometers with which observations were made daily during the cold weather referred to. I take the following from Balmat's register.

Date.	Minimum. temperature Centigrade.	Date.	Minimum. temperature Centigrade.
December 16	− 15°	December 23	− 4½
„ 17	− 20	„ 24	− 6½
„ 18	− 16½	„ 25	− 2
„ 19	− 9	„ 26	+ 2
„ 20	− 13	„ 27	− 3
„ 21	− 20½	„ 28	− 10½
„ 22	− 4¼	„ 29	− 6

The temperature at the Montanvert during the above period may be assumed as generally some degrees lower, so that for a considerable period, previous to my winter observations, the portion of the Mer de Glace near the Montanvert had been exposed to a very low temperature. I reached the place after the weather had become warm, but during my stay there the maximum temperature did not exceed $-4\frac{1}{2}°$ C. Considering therefore the long drain to which the glacier had been subjected previous to the 29th of December, it is not unreasonable to infer that the capillary supply assumed by Professor Forbes must by that time have been exhausted. Notwithstanding this, the motion of the glacier at the Montanvert amounted at the end of December to half its maximum summer motion.

The observations of Balmat which have been published by Professor Forbes* also militate, as far as they go, against the idea of proportionality between the capillary supply and the motion. If the temperatures recorded apply to the Mer de Glace during the periods of observation, it would follow that from the 19th of December 1846 to the 12th of April 1847 the temperature of the air was constantly under zero Centigrade, and hence, during this time, the gorging of the capillaries, which is due to superficial melting, must have ceased. Still, throughout this entire period of depletion the motion of the glacier steadily increased from twenty-four inches to thirty-four and a half inches a day. What has been here said of the Montanvert, and of the points lower down where Balmat's measurements were made, of course applies with greater force to the higher portions of the glacier, which are withdrawn from the operation of superficial melting for a longer period, and which, nevertheless, if I understand Professor Forbes aright, have their motion *least affected* in winter. He records, for example, an observation of Mr. Bakewell's, by which the Glacier des Bossons is shown to be stationary at its end, while its upper portions are moving at the rate of a foot a day. This surely indicates that, at those places where the glacier is longest cut off from superficial supply, the motion is least reduced, which would be a most strange

* 'Occ. Pap.,' p. 224.

Q

result if the motion depended, as affirmed, upon the gorging of the capillaries.

The perusal of the conclusion of Professor Forbes's last volume shows me that a thought similar to that expressed above occurred to Mr. Bakewell also. Speaking of a shallow glacier which moved when the alleged tempera- ture was so enormously below the freezing point that Professor Forbes regards it as unlikely (in which I agree with him), Mr. Bakewell asks, "Is it possible that infil- trated water can have any action whatever under such cir- cumstances?" The reply of Professor Forbes contains these words :— "I have nowhere affirmed the presence of liquid water to be a *sine quâ non* to the plastic motion of glaciers." This statement, I confess, took me by surprise, which was not diminished by further reading. Speaking of the influence of temperature on the motion of the Mer de Glace, Pro- fessor Forbes says, the glacier "took no real start until the frost had given way, and the tumultuous course of the Arveiron showed that its veins were again filled with the circulating medium to which the glacier, like the organic frame, owes its moving energy." * And again :—"It is this fragility precisely which, yielding to the hydrostatic pres- sure of the unfrozen water contained in the countless capil- laries of the glacier, produces the crushing action which shoves the ice over its neighbour particles."

After the perusal of the foregoing paragraphs the reader will probably be less interested in the question as to whether the assumed capillaries exist at all in the glacier. According to Mr. Huxley's observations, they do not.† During the summer of 1857 he carefully experimented with coloured liquids on the Mer de Glace and its tribu- taries, and in no case was he able to discover these fissures in the sound unweathered ice. I have myself seen the red

* Reprint of 'Memoirs in Phil. Trans.' 'Occ. Pap.,' p. 138.
† 'Occ. Pap.,' p. 47.

liquid resting in an auger-hole, where it had lain for an hour without diffusing itself in any sensible degree. This cavity intersected both the white ice and the blue veins of the glacier; and Mr. Huxley, in my presence, cut away the ice until the walls of the cavity became extremely thin, still no trace of liquid passed through them. Experiments were also made upon the higher portions of the Mer de Glace, and also on the Glacier du Géant, with the same result. Thus the very existence of these capillaries is rendered so questionable, that no theory of glacier-motion which invokes their aid could be considered satisfactory.

THOMSON'S THEORY.

(21.)

In the 'Transactions' of the Royal Society of Edinburgh for 1849 is published a very interesting paper by Prof. James Thomson of Queen's College, Belfast, wherein he deduces, as a consequence of a principle announced by the French philosopher Carnot, that water, when subjected to pressure, requires a greater cold to freeze it than when the pressure is removed. He inferred that the lowering of the freezing point for every atmosphere of pressure amounted to .0075 of a degree Centigrade. This deduction was afterwards submitted to the test of experiment by his distinguished brother Prof. Wm. Thomson, and proved correct. On the fact thus established is founded Mr. James Thomson's theory of the "Plasticity of Ice as manifested in Glaciers."

The theory is this :—Certain portions of the glacier are supposed first to be subjected to pressure. This pressure liquefies the ice, the water thus produced being squeezed through the glacier in the direction in which it can most easily escape. But cold has been evolved by the act of liquefaction, and, when the water has been relieved from the pressure, it freezes in a new position. The pressure being thus abolished at the place where it was first applied, new portions of the ice are subjected to the force; these in their turn liquefy, the water is dispersed as before, and re-frozen in some other place. To the succession of processes here assumed Mr. Thomson ascribes the changes of form observed in glaciers.

This theory was first communicated to the Royal Society through the author's brother, Prof. William Thomson, and is printed in the 'Proceedings' of the Society for May, 1857. It was afterwards communicated to the British Association in Dublin, in whose 'Reports' it is further published; and again it was communicated to the Belfast Literary and Philosophical Society, in whose 'Proceedings' it also finds a place.

On the 24th of November, 1859, Mr. James Thomson communicated to the Royal Society, through his brother, a second paper, in which he again draws attention to his theory. He offers it in substitution for my views as the best argument that he can adduce against them; he also controverts the explanations of regelation propounded by Prof. James D. Forbes and Prof. Faraday, believing that his own theory explains all the facts so well as to leave room for no other.

But the passage in this paper which demands my chief attention is the following:—"Prof. Tyndall (writes Mr. Thomson), in papers and lectures subsequent to the publication of this theory, appears to adopt it to some extent, and to endeavour to make its principles co-operate with the views he had previously founded on Mr. Faraday's fact of regelation." I may say that Mr. Thomson's main thought was familiar to me long before his first communication on the plasticity of ice appeared; but it had little influence upon my convictions. Were the above passage correct, I should deserve censure for neglecting to express my obligations far more explicitly than I have hitherto done; but I confess that even now I do not understand the essential point of Mr. Thomson's theory,—that is to say, its application to the phenomena of glacier motion. Indeed, it was the obscurity in my mind in connexion with this point, and the hope that time might enable me to seize more clearly upon his meaning, which prevented me from giving that prominence to the theory of Mr. Thomson which, for aught I know, it may well deserve. I will here briefly state one or two of my difficulties, and shall feel very grateful to have them removed.

Let us fix our attention on a vertical slice of ice transverse to the glacier, and to which the pressure is applied perpendicular to its surfaces. The ice liquefies, and, supposing the means of escape offered to the compressed water to be equal all round, it is plain that there will be as great a tendency to squeeze the water upwards as downwards; for the mere tendency to flow down by its own gravity becomes, in comparison to the forces here acting on the water, a vanishing quantity. But the fact is, that the ice *above* the slice is more permeable than that below it; for, as we descend a glacier, the ice becomes more compact. Hence the greater part of the dispersed water will be refrozen on that side of the slice which is turned towards the origin of the glacier; and the consequence is, that, according to Mr. Thomson's principle, the glacier ought to move up hill instead of down.

I would invite Mr. Thomson to imagine himself and me together upon the ice, desirous of examining this question in a philosophic spirit; and that we have taken our places beside a stake driven into the ice, and descending with the glacier. We watch the ice surrounding the stake, and find that every speck of dirt upon it retains its position; there is no liquefaction of the ice that bears the dirt, and consequently it rests on the glacier undisturbed. After twelve hours we find the stake fifteen inches distant from its first position: I would ask Mr. Thomson how did it get there? Or let us fix our attention on those six stakes which M. Agassiz drove into the glacier of the Aar in 1841, and found erect in 1842 at some hundreds of feet from their first position:—how did they get there? How, in fine, does the end of a glacier become its end? Has it been liquefied and re-frozen? If not, it must have been *pushed* down by the very forces which Mr. Thomson invokes to produce his liquefaction. Both the liquefaction, as far as it exists, and the motion, are products of the same cause.

In short, this theory, as it presents itself to my mind, is so powerless to account for the simplest fact of glacier-motion, that I feel disposed to continue to doubt my own competence to understand it rather than ascribe to Mr. Thomson an hypothesis apparently so irrelevant to the facts which it professes to explain.

Another difficulty is the following:—Mr. Thomson will have seen that I have recorded certain winter measurements made on the Mer de Glace, and that these measurements show not only that the ice moves at that period of the year, but that it exhibits those characteristics of motion from which its plasticity has been inferred; the velocity of the central portions of the glacier being in round numbers double the velocity of those near the sides. Had there been any necessity for it, this ratio might have been augmented by placing the side-stakes closer to the walls of the glacier. Considering the extreme coldness of the weather which preceded these measurements, it is a moderate estimate to set down the temperature of the ice in which my stakes were fixed at 5° Cent. below zero.

Let us now endeavour to estimate the pressure existing at the portion of the glacier where these measurements were made. The height of the Montanvert above the sea-level is, according to Prof. Forbes, 6300 feet; that of the Col du Géant, which is the summit of the principal tributary of the Mer de Glace, is 11,146 feet: deducting the former from the latter, we find the height of the Col du Géant above the Montanvert to be 4846 feet.

Now, according to Mr. Thomson's theory and his brother's experiments, the melting point of ice is lowered .0075 Centigrade for every atmosphere of pressure; and one atmosphere being equivalent to the pressure of about thirty-three feet of water, we shall not be over the truth if we take the height of an equivalent column of glacier-ice, of a compactness the mean of those which it exhibits upon the Col du Géant and at the Montanvert respectively, at forty feet. The compactness of glacier ice is, of course, affected by the air-bubbles contained within it.

If, then, the pressure of forty feet of ice lower the melting point .0075 Centigrade, it follows that the pressure of a column 4846 feet high will lower it nine-tenths of a degree Centigrade. Supposing, then, *the unimpeded thrust of the whole glacier, from the Col du Géant downwards,* to be exerted on the ice at the Montanvert; or, in other words, supposing the bed of the glacier to be absolutely smooth and every trace of friction abolished, the utmost the pressure thus obtained could perform would be to lower the melting point of the Montanvert ice by the quantity above mentioned. Taking into account the actual state of things, the friction of the glacier against its sides and bed, the opposition which the three tributaries encounter in the neck of the valley at Trélaporte, the resistance encountered in the sinuous valley through which it passes; and finally, bearing in mind the comparatively short length of the glacier, which has to bear the thrust, and oppose the latter by its own friction merely;—I think it will appear evident that the ice at the Montanvert cannot possibly have its melting point lowered by pressure more than a small fraction of a degree.

The ice in which my stakes were fixed being −5° Centigrade, according to Mr. Thomson's calculation and his brother's experiments, it would require 667 atmospheres of pressure to liquefy it; in other words, it would require the unimpeded pressure of a column of glacier-ice 26,680 feet high. Did Mont Blanc rise to two and a half times its present height above the Montanvert, and were the latter place connected with the summit of the mountain by a continuous glacier with its bed absolutely smooth, the pressure at the Montanvert would be rather under that necessary to liquefy the ice on which my winter observations were made.

If it be urged that, though the temperature near the surface may be several degrees below the freezing point, the great body of the glacier does not share this temperature, but is, in all probability, near to 32°, my reply is simple. I did not measure the motion of the ice in the body of the glacier; nobody ever did; my measurements refer to the ice at and near the surface, and it is this ice which showed the plastic deportment which the measurements reveal.

Such, then, are some of the considerations which prevent me from accepting the theory of Mr. Thomson, and I trust they will acquit me of all desire to make his theory co-operate with my views. I am, however, far from considering his deduction the less important because of its failing to account for the phenomena of glacier motion.

THE PRESSURE-THEORY OF GLACIER-MOTION.

(22.)

BROADLY considered, two classes of facts are presented to
the glacier-observer; the one suggestive of viscosity, and
the other of the reverse.   The former are seen where *pressure* comes into play, the latter where *tension* is operative.
By pressure ice can be moulded to any shape, while the
same ice snaps sharply asunder if subjected to tension.
Were the result worth the labour, ice might be moulded
into vases, statuettes, bent into spiral bars, and, I doubt
not, by the proper application of pressure, a *rope* of ice
might be formed and coiled into a *knot*.   But not one of
these experiments, though they might be a thousand-fold
more striking than any ever made upon a glacier, would
in the least demonstrate that ice is really a viscous body.

I have here stated what I believe to be feasible.   Let
me now refer to the experiments which have been actually

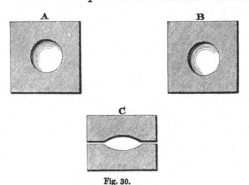

Fig. 30.

made in illustration of this point.   Two pieces of seasoned
box-wood had corresponding cavities hollowed in them, so
that, when one was placed upon the other, a lenticular

space was enclosed. A and B, Fig. 30, represent the pieces of box-wood with the cavities in plan: C represents their section when they are placed upon each other.

A *sphere* of ice rather more than sufficient to fill the lenticular space was placed between the pieces of wood and subjected to the action of a small hydraulic press. The ice was crushed, but the crushed fragments soon re-attached themselves, and, in a few seconds, a lens of compact ice was taken from the mould.

This lens was placed in a cylindrical cavity hollowed out in another piece of box-wood, and represented at C, Fig. 31; and a flat piece of the wood was placed over the lens as a cover, as at D. On subjecting the whole to pressure, the lens broke, as the sphere had done, but the crushed mass soon re-established its continuity, and in less than half a minute a compact cake of ice was taken from the mould.

Fig. 31.

In the following experiment the ice was subjected to a still severer test:—A hemispherical cavity was formed in one block of box-wood, and upon a second block a hemispherical protuberance was turned, smaller than the cavity, so that, when the latter was placed in the former, a space of a quarter of an inch existed between both. Fig. 32 represents a section of the two pieces of box-wood; the brass pins *a*, *b*, fixed in the slab G H, and entering suitable apertures in the mould I K, being intended to keep the two surfaces concentric. A lump of ice being placed in

Fig. 32.

the cavity, the protuberance was brought down upon it, and the mould subjected to hydraulic pressure: after a short interval the ice was taken from the mould as a smooth compact *cup,* its crushed particles having reunited, and established their continuity.

To make these results more applicable to the bending of glacier-ice, the following experiments were made:—A block of box-wood, M, Fig. 33, 4 inches long, 3 wide, and 3 deep,

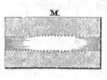

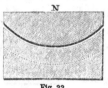

Fig. 33.

had its upper surface slightly curved, and a groove an inch wide, and about an inch deep, worked into it. A corresponding plate was prepared, having its under surface part of a convex cylinder, of the same curvature as the concave surface of the former piece. When the one slab was placed upon the other, they presented the appearance represented in section at N. A straight prism of ice 4 inches long, an inch wide, and a little more than an inch in depth, was placed in the groove; the upper slab was placed upon it, and the whole was subjected to the hydraulic press. The prism broke, but, the quantity of ice being rather more than sufficient to fill the groove, the pressure soon brought the fragments together and re-established the continuity of the ice. After a few seconds it was taken from the mould a bent bar of ice. This bar was afterwards passed through three other moulds of gradually augmenting curvature, and was taken from the last of them a *semi-ring* of compact ice.

The ice, in changing its form from that of one mould to that of another, was in every instance broken and crushed by the pressure; but suppose that instead of three moulds three thousand had been used; or, better still, suppose the curvature of a single mould to change by extremely slow degrees; the ice would then so gradually change its form that no rude rupture would be apparent. Practically the ice would behave as a *plastic* substance; and indeed this plasticity has been contended for by M. Agassiz, in opposition to the idea of viscosity. As already stated, the ice, bruised, and flattened, and bent in the above experiments, was incapable of being sensibly stretched; it was plastic to pressure but not to tension.

A quantity of water was always squeezed out of the crushed ice in the above experiments, and the bruised fragments were intermixed with this and with air. Minute quantities of both remained in the moulded ice, and thus rendered it in some degree turbid. Its character, however, as to continuity may be inferred from the fact that the ice-cup, moulded as described, held water without the slightest visible leakage.

Ice at 32° may, as already stated, be crushed with extreme facility, and glacier-ice with still more readiness than lake-ice: it may also be scraped with a knife with even greater facility than some kinds of chalk. In comparison with ice at 100° below the freezing point, it might be popularly called *soft*. But its softness is not that of paste, or wax, or treacle, or lava, or honey, or tar. It is the softness of calcareous spar in comparison with that of rock-crystal; and although the latter is incomparably harder than the former, I think it will be conceded that the term viscous would be equally inapplicable to both. My object here is clearly to define terms, and not permit physical error to lurk beneath them. How far this ice, with a softness thus defined, when subjected to the gradual pressures exerted in a glacier, is bruised and broken, and how far the motion of its parts may approach to that of a truly viscous body under pressure, I do not know. The critical point here is that the ice changes its form, and preserves its continuity, during its motion, in virtue of *external* force. It remains continuous whilst it moves, because its particles are kept in juxta-position by pressure, and when this external prop is removed, and the ice, subjected to tension, has to depend solely upon the mobility of its own particles to preserve its continuity, the analogy with a viscous body instantly breaks down.*

* "Imagine," writes Professor Forbes, "a long narrow trough or canal, stopped at both ends and filled to a considerable depth with treacle, honey, tar, or any such viscid fluid. Imagine one end of the trough to give way, the bottom still remaining horizontal: if the friction of the fluid against the bottom be greater than the friction against its own particles, the upper strata will roll over the lower ones, and protrude in a convex slope, which will be propagated backwards towards the other or closed end of the trough. Had the matter been quite fluid the whole would have run out, and spread itself on a level: as it is, it assumes precisely the conditions which we suppose to exist in a glacier." This is perfectly definite, and my equally definite opinion is that no glacier ever exhibited the mechanical effects implied by this experiment.

REGELATION.

(23.)

I WAS led to the foregoing results by reflecting on an experiment performed by Mr. Faraday, at a Friday evening meeting of the Royal Institution, on the 7th of June, 1850, and described in the 'Athenæum' and 'Literary Gazette' for the same month. Mr. Faraday then showed that when two pieces of ice, with moistened surfaces, were placed in contact, they became cemented together by the freezing of the film of water between them, while, when the ice was below 32° Fahr., and therefore *dry*, no effect of the kind could be produced. The freezing was also found to take place under water; and indeed it occurs even when the water in which the ice is plunged is as hot as the hand can bear.

A generalisation from this interesting fact led me to conclude that a bruised mass of ice, if closely confined, must re-cement itself when its particles are brought into contact by pressure; in fact, the whole of the experiments above recorded immediately suggested themselves to my mind as natural deductions from the principle established by Faraday. A rough preliminary experiment assured me that the deductions would stand testing; and the construction of the box-wood-moulds was the consequence. We could doubtless mould many solid substances to any extent by suitable pressure, breaking the attachment of their particles, and re-establishing a certain continuity by the mere force of cohesion. With such substances, to which we should never think of applying the term viscous, we might also imitate the changes of form to which glaciers are subject: but, superadded to the mere cohesion which here comes into play, we have, in the case of

ice, the actual regelation of the severed surfaces, and consequently a more perfect solid.   In the Introduction to this
book I have referred to the production of slaty cleavage by
pressure; and at a future page I hope to show that the
lamination of the ice of glaciers is due to the same cause;
but, as justly observed by Mr. John Ball, there is no tendency to cleave in the *sound* ice of glaciers; in fact, this
tendency is obliterated by the perfect regelation of the
severed surfaces.

Mr. Faraday has recently placed pieces of ice, in water,
under the strain of forces tending to pull them apart.
When two such pieces touch at a single point they adhere
and move together as a rigid piece; but a little lateral
force carefully applied breaks up this union with a crackling noise, and a new adhesion occurs which holds the
pieces together in opposition to the force which tends to
divide them.   Mr. James Thomson had referred regelation
to the cold produced by the liquefaction of the pressed ice;
but in the above experiment all pressure is not only taken
away, but is replaced by tension.   Mr. Thomson also con-
ceives that, when pieces of ice are simply placed together
without intentional pressure, the capillary attraction brings
the pressure of the atmosphere into play; but Mr. Faraday
finds that regelation takes place in *vacuo*.   A true viscidity
on the part of ice Mr. Faraday never has observed, and he
considers that his recent experiments support the view
originally propounded by himself, namely, that a particle
of water on a surface of ice becomes solid when placed
between two surfaces, because of the increased influence
due to their joint action.

CRYSTALLIZATION AND INTERNAL LIQUE-FACTION.

(24.)

In the Introduction to this book I have briefly referred to the force of crystallization. To permit this force to exercise its full influence, it must have free and unimpeded action; a crystal, for instance, to be properly built, ought to be suspended in the middle of the crystallizing solution, so that the little architects can work all round it; or if placed upon the bottom of a vessel, it ought to be frequently turned, so that all its facets may be successively subjected to the building process. In this way crystals can be *nursed* to an enormous size. But where other forces mingle with that of crystallization, this harmony of action is destroyed; the figures, for example, that we see upon a glass window, on a frosty morning, are due to an action compounded of the pure crystalline force and the cohesion of the liquid to the window-pane. A more regular effect is obtained when the freezing particles are suspended in still air, and here they build themselves into those wonderful figures which Dr. Scoresby has observed in the Polar Regions, Mr. Glaisher at Greenwich, and I myself on the summit of Monte Rosa and elsewhere.

Not only however in air, but in water also, figures of great beauty are sometimes formed. Harrison's excellent machine for the production of artificial ice is, I suppose, now well known; the freezing being effected by carrying brine, which had been cooled by the evaporation of ether, round a series of flat tin vessels containing water. The latter gradually freezes, and, on watching those vessels while the action was proceeding very slowly, I have seen little six-rayed stars of thin ice forming, and rising to the surface of the liquid. I believe the fact was never before observed, but it would be interesting to follow it up, and to develop experimentally this most interesting case of crystallization.

The surface of a freezing lake presents to the eye of the observer nothing which could lead him to suppose that a similar molecular architecture is going on there. Still the particles are undoubtedly related to each other in this way; they are arranged together on this starry type. And not only is this the case at the surface, but the largest blocks of ice which reach us from Norway and the Wenham Lake are wholly built up in this way. We can reveal the internal constitution of these masses by a reverse process to that which formed them; we can send an agent into the interior of a mass of ice which shall take down the atoms which the crystallizing forces had set up. This agent is a solar beam; with which it first occurred to me to make this simple experiment in the autumn of 1857. I placed a large converging lens in the sunbeams passing through a room, and observed the place where the rays were brought to a focus behind the lens; then shading the lens, I placed a clear cube of ice so that the point of convergence of the rays might fall within it. On removing the screen from the lens. a cone of sunlight went through the cube, and along the course of the cone the ice became studded with lustrous spots, evidently formed by the beam, as if minute reflectors had been suddenly established within the mass, from which the light flashed when it met them. On examining the cube afterwards I found that each of these spots was surrounded by a liquid flower of six petals; such flowers were distributed in hundreds through the ice, being usually clear and detached from each other, but sometimes crowded together into liquid bouquets, through which, however, the six-starred element could be plainly traced. At first the edges of the

leaves were unbroken curves, but when the flowers expanded
under a long-continued action, the edges became serrated.
When the ice was held at a suitable angle to the solar
beams, these liquid blossoms, with their central spots
shining more intensely than burnished silver, presented
an exhibition of beauty not easily described. I have given
a sketch of their appearance in Fig. 34.

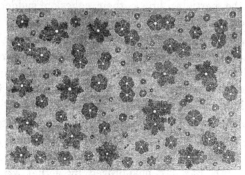

Fig. 34.

I have here to direct attention to an extremely curious
fact. On sending the sunbeam through the transparent
ice, I often noticed that the appearance of the lustrous
spots was accompanied by an audible clink, as if the ice
were ruptured inwardly. But there is no ground for
assuming such rupture, and on the closest examination
no flaw is exhibited by the ice. What then can be the
cause of the noise? I believe the following considerations
will answer the question:—

Water always holds a quantity of air in solution, the
diffusion of which through the liquid, as proved by M.
Donny, has an immense effect in weakening the cohesion
of its particles; recent experiments of my own show that
this is also the case in an eminent degree with many
volatile liquids. M. Donny has proved that, if water be
thoroughly purged of its air, a long glass tube filled with

this liquid may be inverted, while the tenacity with which the water clings to the tube, and with which its particles cling to each other, is so great that it will remain securely suspended, though no external hindrance be offered to its descent. Owing to the same cause, water deprived of its air will not boil at 212° Fahr., and may be raised to a temperature of nearly 300° without boiling; but when this occurs the particles break their cohesion suddenly, and ebullition is converted into explosion.

Now, when ice is formed, every trace of the air which the water contained is squeezed out of it; the particles in crystallizing reject all extraneous matter, so that in ice we have a substance quite free from the air, which is never absent in the case of water; it therefore follows that if we could preserve the water derived from the melting of ice from contact with the atmosphere, we should have a liquid eminently calculated to show the effects described by M. Donny. Mr. Faraday has proved by actual experiment that this is the case.

Let us apply these facts to the explanation of the clink heard in my experiments. On sending a sunbeam through ice, liquid cavities are suddenly formed at various points within the mass, and these cavities are completely cut off from atmospheric contact. But the water formed by the melting ice is less in volume than the ice which produces it; the water of a cavity is not able to fill it, hence a vacuous space must be formed in the cell. I have no doubt that, for a time, the strong cohesion between the walls of the cell and the drop within it augments the volume of the latter a little, so as to compel it to fill the cell; but as the quantity of liquid becomes greater the shrinking force augments, until finally the particles snap asunder like a broken spring. At the same moment a lustrous spot appears, which is a vacuum, and simultaneously with the appearance of this vacuum the clink

was always heard. Multitudes of such little explosions
must be heard upon a glacier when the strong summer
sun shines upon it, the aggregate of which must, I think,
contribute to produce the "crepitation" noticed by M.
Agassiz, and to which I have already referred.

In Plate VI. of the Atlas which accompanies the 'Système Glaciaire' of M. Agassiz, I notice drawings of figures
like those I have described, which he has observed in glacier-
ice, and which were doubtless produced by direct solar radiation. I have often myself observed figures of exquisite
beauty formed in the ice on the surface of glacier-pools by
the morning sun. In some cases the spaces between the
leaves of the liquid flowers melt partially away, and leave
the central spot surrounded by a crimped border; sometimes
these spaces wholly disappear, and the entire space bounded
by the lines drawn from point to point of the leaves becomes liquid, thus forming perfect hexagons. The crimped
borders exhibit different degrees of serration, from the
full leaves themselves to a gentle undulating line, which
latter sometimes merges into a perfect circle. In the ice
of glaciers, I have seen the internal liquefaction ramify
itself like sprigs of myrtle; in the same ice, and particularly towards the extremities of the glacier, disks innumerable are also formed, consisting of flat round liquid
spaces, a bright spot being usually associated with each.
These spots have been hitherto mistaken for air-bubbles;
but both they and the lustrous disks at the centres of the
flowers are vacuous. I proved them to be so by plunging
the ice containing them into hot water, and watching
what occurred when the walls of the cells were dissolved,
and a liquid connexion established between them and the
atmosphere. In all cases they totally collapsed, and no
trace of air rose to the surface of the warm water.

No matter in what direction a solar beam is sent through
lake-ice, the liquid flowers are all formed parallel to the

surface of freezing. The beam may be sent parallel, perpendicular, or oblique to this surface; the flowers are always formed in the same planes. Every line perpendicular to the surface of a frozen lake is in fact an axis of symmetry, round which the molecules so arrange themselves, that, when taken down by the delicate fingers of the sunbeam, the six-leaved liquid flowers are the result.

In the ice of glaciers we have no definite planes of freezing. It is first snow, which has been disturbed by winds while falling, and whirled and tossed about by the same agency after it has fallen, being often melted, saturated with its own water, and refrozen: it is cast in shattered fragments down cascades, and reconsolidated by pressure at the bottom. In ice so formed and subjected to such mutations, definite planes of freezing are, of course, out of the question.

The flat round disks and vacuous spots to which I have referred come here to our aid, and furnish us with an entirely new means of analysing the internal constitution of a glacier. When we examine a mass of glacier-ice which contains these disks, we find them lying in all imaginable planes; not confusedly, however—closer examination shows us that the disks are arranged in groups, the members of each group being parallel to a common plane, but the parallelism ceases when different groups are compared. The effect is exactly what would be observed, supposing ordinary lake-ice to be broken up, shaken together, and the confused fragments regelated to a compact continuous mass. In such a jumble the original planes of freezing would lie in various directions; but no matter how compact or how transparent ice thus constituted might appear, a solar beam would at once reveal its internal constitution by developing the flowers parallel to the planes of freezing of the respective fragments. A sunbeam sent through glacier-ice always reveals the flowers in the planes of the disks, so that the latter alone at once informs us of its crystalline constitution.

Hitherto, as I have said, these disks have been mistaken for bubbles containing air, and their flattening has been ascribed to the pressure to which they have been subjected. M. Agassiz thus refers to them :—"The air-bubbles undergo no less curious modifications. In the neighbourhood of the *névé*, where they are most numerous, those which one sees on the surface are all spherical or ovoid, but by degrees they begin to be flattened, and near the end of the glacier there are some that are so flat *that they might be taken for fissures when seen in profile*. The drawing represents a piece of ice detached from the gallery of infiltration. All the bubbles are greatly flattened. But what is most extraordinary is, that, far from being uniform, *the flattening is different in each fragment;* so that the bubbles, according to the face which they offer, appear either very broad or very thin." This description of glacier-ice is correct: it agrees with the statements of all other observers. But there are two assumptions in the description which must henceforth be given up; first, the bubbles seen like fissures in profile are not air-bubbles at all, but vacuous spots, which the very constitution of ice renders a necessary concomitant of its inward melting; secondly, the assumption that the bubbles have been *flattened* by pressure must be abandoned; for they are found, and may be developed at will, in lake-ice on which no pressure has been exerted.

But these remarks dispose only of a certain class of cells contained in glacier ice. Besides the liquid disks and vacuous spots, there are innumerable true bubbles entangled in the mass. These have also been observed and described by M. Agassiz; and Mr. Huxley has also given us an accurate account of them. M. Agassiz frequently found air and water associated in the same cell. Mr.

Huxley found no exception to the rule: in each case the bubble of air was enclosed in a cell which was also partially filled with water. He supposes that the water may be that of the originally-melted snow which has been carried down from the *névé* unfrozen. This hypothesis is worthy of a great deal more consideration than I have had time to give to it, and I state it here in the hope that it will be duly examined.

My own experience of these associated air and water cells is derived almost exclusively from lake-ice, in which I have often observed them in considerable numbers. In examining whether the liquid contents had ever been frozen or not, I was guided by the following considerations. If the air be that originally entangled in the solid, it will have the ordinary atmospheric density at least; but if it be due to the melting of the walls of the cell, then the water so formed being only eight-ninths of that of the ice which produced it, *the air of the bubble must be rarefied.* I suppose I have made a hundred different experiments upon these bubbles to determine whether the air was rarefied or not, and in every case found it so. Ice containing the bubbles was immersed in warm water, and always, when the rigid envelope surrounding a bubble was melted away, the air suddenly collapsed to a fraction of its original dimensions. I think I may safely affirm that, in some cases, the collapse reduced the bubbles to the thousandth part of their original volume. From these experiments I should undoubtedly infer, that in lake-ice at least, the liquid of the cells is produced by the melting of the ice surrounding the bubbles of air.

But I have not subjected the bubbles of glacier-ice to the same searching examination. I have tried whether the insertion of a pin would produce the collapse of the bubbles, but it did not appear to do so. I also made a few experiments at Rosenlaui, with warm water, but the result was not satisfactory. That ice melts internally at the surfaces of the bubbles is, I think, rendered certain by my experiments, but whether the water-cells of glacier-ice are entirely due to such melting, subsequent observers will no doubt determine.

I have found these composite bubbles at all parts of glaciers; in the ice of the moraines, over which a protective covering had been thrown; in the ice of sand-cones, after the removal of the superincumbent débris; also in ice taken from the roofs of caverns formed in the glacier, and which the direct sun-light could hardly by any possibility attain. That ice should liquefy at the surface of a cavity is, I think, in conformity with all we know concerning the physical nature of heat. Regarding it as a motion of the particles, it is easy to see that this motion is less restrained at the surface of a cavity than in the solid itself, where the oscillation of each atom is controlled by the particles which surround it; hence *liquid liberty*, if I may use the term, is first attained at the surface. Indeed I have proved by experiment that ice may be melted internally by heat which has been conducted through its external portions without melting them. These facts are the exact complements of those of "regelation;" for here, two moist surfaces of ice being brought into close contact, their liquid liberty is destroyed and the surfaces freeze together.

R

THE MOULINS.

(25.)

THE first time I had an opportunity of seeing these remarkable glacier-chimneys was in the summer of 1856, upon the lower glacier of Grindelwald. Mr. Huxley was my companion at the time, and on crossing the so-called Eismeer we heard a sound resembling the rumble of distant thunder, which proceeded from a perpendicular shaft formed in the ice, and into which a resounding cataract discharged itself. The tube in fact resembled a vast organ-pipe, whose thunder-notes were awakened by the concussion of the falling water, instead of by the gentle flow of a current of air. Beside the shaft our guide hewed steps, on which we stood in succession, and looked into the tremendous hole. Near the first shaft was a second and smaller one, the significance of which I did not then understand ; it was not more than 20 feet deep, but seemed filled with a liquid of exquisite blue, the colour being really due to the magical shimmer from the walls of the moulin, which was quite empty. As far as we could see, the large shaft was vertical, but on dropping a stone into it a shock was soon heard, and after a succession of bumps, which occupied in all seven seconds, we heard the stone no more. The depth of the moulin could not be thus ascertained, but we soon found a second and still larger one which gave us better data. A stone dropped into this descended without interruption for four seconds, when a concussion was heard ; and three seconds afterwards the final shock was audible : there was thus but a single interruption in the descent. Supposing all the acquired velocity to have been destroyed by the shock, by adding the space passed over by the stone in four and in three seconds respectively, and making allowance for the time required by the sound to ascend from the bottom, we find the depth of the shaft to be about 345 feet. There is, however, no reason to suppose that this measures the depth of the glacier at the place referred to. These shafts are to be found in almost all great glaciers; they are very numerous in the Unteraar glacier, numbers of them however being empty. On the Mer de Glace they are always to be found in the region of Trélaporte, one of the shafts there being, *par excellence*, called the Grand Moulin. Many of them also occur on the Glacier de Léchaud.

As truly observed by M. Agassiz, these moulins occur only at those parts of the glacier which are not much rent by fissures, for only at such portions can the little rills produced by superficial melting collect to form streams of any magnitude. The valley of unbroken ice formed in the Mer de Glace near Trélaporte is peculiarly favourable for the collection of such streams; we see the little rills commencing, and enlarging by the contributions of others, the trunk-rill pouring its contents into a little stream which stretches out a hundred similar arms over the surface of the glacier. Several such streams join, and finally a considerable brook, which receives the superficial drainage of a large area, cuts its way through the ice.

But although this portion of the glacier is free from those long-continued and permanent strains which, having once rent the ice, tend subsequently to widen the rent and produce yawning crevasses, it is not free from local strains sufficient to produce *cracks* which penetrate the glacier to a great depth. Imagine such a crack intersecting such a glacier-rivulet as we have described. The water rushes down it, and soon scoops a funnel large enough to engulf the entire stream. The moulin is thus formed, and, as the ice moves downward, the sides of the crack are squeezed together and regelated, the seam which marks the line of junction being in most cases distinctly visible. But as the motion continues, other portions of the glacier come into the same state of strain as that which produced the first crack; a second one is formed across the stream, the old shaft is forsaken, and a new one is hollowed out, in which for a season the cataract plays the thunderer. I have in some cases counted the forsaken shafts of six old moulins in advance of an active one. Not far from the Grand Moulin of the Mer de Glace in 1857 there was a second empty shaft, which evidently communicated by a subglacial duct with that into which the torrent was precipitated. Out of the old orifice issued a strong cold blast, the air being manifestly impelled through the duct by the falling water of the adjacent moulin.

These shafts are always found in the same locality; the portion of the Mer de Glace to which I have referred is never without them. Some of the guides affirm that they are motionless; and a statement of Prof. Forbes has led to the belief that this was also his opinion.* M. Agassiz, however, observed the motion of some of these shafts upon the glacier of the Aar; and when on the spot in 1857, I was anxious to decide the point by accurate measurements with the theodolite.

My friend Mr. Hirst took charge of the instrument, and on the 28th of July, I fixed a single stake beside the Grand Moulin, in a straight line between a station at Trélaporte and a well-defined mark on the rock at the opposite side of the valley. On the 31st, the displacement of the stake amounted to 50 inches, and on the 1st of August it had moved 74½ inches—the moulin, to all appearance, occupying throughout the same position with regard to the stake.

* "Every year, and year after year, the watercourses follow the same lines of direction—their streams are precipitated into the heart of the glacier by vertical funnels, called 'moulins,' at the very same points."— Forbes's Fourth Letter upon Glaciers: 'Occ. Pap.', p. 29.

To render this certain, moreover we subsequently drove two additional stakes into the ice, thus enclosing the mouth of the shaft in a triangle.   On the 8th of August the displacements were measured and gave the following results :—

	Total Motion.
First (old) stake	198 inches.
Second (new) do.	123 „
Third	124 „

The old stake had been fixed for 11 days, and its daily motion—*which was also that of the moulin*—averaged 18 inches a-day.   Hence the moulins share the general motion of the glacier, and their apparent permanence is not, as has been alleged, a proof of the semi-fluidity of the glacier, but is due to the breaking of the ice as it passes the place of local strain.

Wishing to obtain some estimate as to the depth of the ice, Mr. Hirst undertook the sounding of some of the moulins upon the Glacier de Léchaud, making use of a tin vessel filled with lumps of lead and iron as a weight. The cord gave way and he lost his plummet.   To measure the depth of the Grand Moulin, we obtained fresh cord from Chamouni, to which we attached a four-pound weight. Into a cavity at the bottom of the weight we stuffed a quantity of butter, to indicate the nature of the bottom against which the weight might strike.   The weight was dropped into the shaft, and the cord paid out until its slackening informed us that the weight had come to rest; by shaking the string, however, and walking round the edge of the shaft, the weight was liberated, and sank some distance further.   The cord partially slackened a second time, but the strain still remaining was sufficient to render it doubtful whether it was the weight or the action of the falling water which produced it.   We accordingly paid out the cord to the end, but, on withdrawing it, found that the greater part of it had been coiled and knotted up by the falling water.

We uncoiled, and sounded again. At a depth of 132 feet the weight reached a ledge or protuberance of ice, and by shaking and lifting it, it was caused to descend 31 feet more. A depth of 163 feet was the utmost we could attain to. We sounded the old moulin to a depth of 90 feet; while a third little shaft, beside the large one, measured only 18 feet in depth. We could see the water escape from it through a lateral canal at its bottom, and doubtless the water of the Grand Moulin found a similar exit. There was no trace of dirt upon the butter, which might have indicated that we had reached the bed of the glacier.

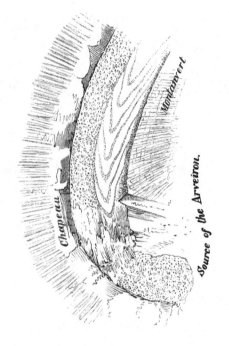

DIRT-BANDS OF THE MER DE GLACE, AS SEEN FROM A POINT
NEAR THE FLEGÈRE.

Fig. 35.   *To face p.* 367.

DIRT-BANDS OF THE MER DE GLACE.

(26.)

THESE bands were first noticed by Prof. Forbes on the 24th of July, 1842, and described by him in the following words:—" My eye was caught by a very peculiar appearance of the surface of the ice, which I was certain that I now saw for the first time. It consisted of nearly hyperbolic brownish bands on the glacier, the curves pointing downwards, and the two branches mingling indiscriminately with the moraines, presenting an appearance of a succession of waves some hundred feet apart." * From no single point of view hitherto attained can all the Dirt-Bands of the Mer de Glace be seen at once. To see those on the terminal portion of the glacier, a station ought to be chosen on the opposite range of the Brévent, a few hundred yards beyond the Croix de la Flegère, where we stand exactly in front of the glacier as it issues into the valley of Chamouni. The appearance of the bands upon the portion here seen is represented in Fig. 35.

It will be seen that the bands are confined to one side of the glacier, and either do not exist, or are obliterated by the débris, upon the other side. The cause of the accumulation of dirt on the right side of the glacier is, that no less than five moraines are crowded together at this side. In the upper portions of the Mer de Glace these moraines are distinct from each other; but in descending, the successive engulfments and disgorgings of the blocks and dirt have broken up the moraines; and at the place now before us the materials which composed them are strewn confusedly on the right side of the glacier. The portion of the ice on which the dirt-bands appear is derived from the Col du

* 'Travels,' page 162.

Géant. They do not quite extend to the end of the glacier, being obliterated by the dislocation of the ice upon the frozen cascade of Des Bois.

Let us now proceed across the valley of Chamouni to the Montanvert; where, climbing the adjacent heights to an elevation of six or eight hundred feet above the hotel, we command a view of the Mer de Glace, from Trélaporte almost to the commencement of the Glacier des Bois. It was from this position Professor Forbes first observed the bands. Fifteen, sixteen, and seventeen years later I observed them from the same position. The number of bands which Professor Forbes counted from this position was eighteen, with which my observations agree. The entire series of bands which I observed, with the exception of one or two, must have been the *successors* of those observed by Professor Forbes; and my finding the same number after an interval of so many years proves that the bands must be due to some regularly recurrent cause. Fig. 36 represents the bands as seen from the heights adjacent to the Montanvert.

I would here direct attention to an analogy between a glacier and a river, which may be observed from the heights above the Montanvert, but to which no reference, as far as I know, has hitherto been made. When a river meets the buttress of a bridge, the water rises against it, and, on sweeping round it, forms an elevated ridge, between which and the pier a depression occurs which varies in depth with the force of the current. This effect is shown by the Mer de Glace on an exaggerated scale. Sweeping round Trélaporte, the ice pushes itself beyond the promontory in an elevated ridge, from which it drops by a gradual slope to the adjacent wall of the valley, thus forming a depression typified by that already alluded to. A similar effect is observed at the opposite side of the glacier on turning round the Echelets; and both combine

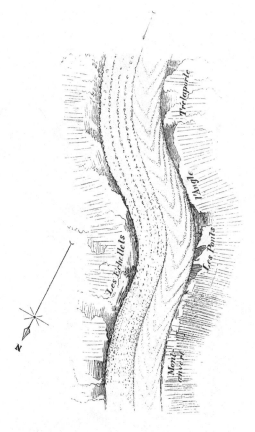

DIRT-BANDS OF THE MER DE GLACE, AS SEEN FROM
LES CHARMOZ.

FIG. 36.   *To face p.* 368.

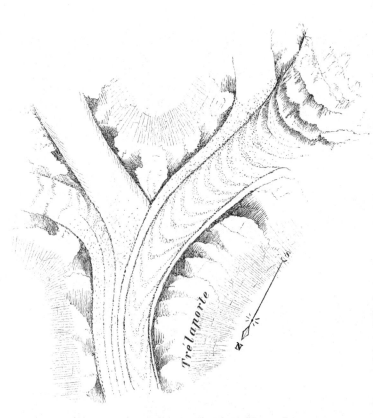

DIRT-BANDS OF THE MER DE GLACE, AS SEEN FROM THE
CLEFT STATION, TRÉLAPORTE.

FIG. 37.   *To face p.* 369.

to form a kind of skew surface. A careful inspection of
the frontispiece will detect this peculiarity in the shape of
the glacier.

From neither of the stations referred to do we obtain
any clue to the origin of the dirt-bands. A stiff but
pleasant climb will place us in that singular cleft in the
cliffy mountain-ridge which is seen to the right of the
frontispiece; and from it we easily attain the high platform of rock immediately to the left of it. We stand here
high above the promontory of Trélaporte, and occupy the
finest station from which the Mer de Glace and its tributaries can be viewed. From this station we trace the dirt-
bands over most of the ice that we have already scanned,
and have the further advantage of being able to follow
them to their very source.

This source is the grand ice-cascade which descends in a
succession of precipices from the plateau of the Col du
Géant into the valley which the Glacier du Géant fills.
We see from our present point of view that the bands
*are confined to the portion of the glacier which has descended
the cascade.* Fig. 37 represents the bands as seen from
the Cleft station above Trélaporte.

We are now however at such a height above the glacier
and at such a distance from the base of the cascade, that we
can form but an imperfect notion of the true contour of the
surface. Let us therefore descend, and walk up the Glacier
du Géant towards the cascade. At first our road is level,
but we gradually find that at certain intervals we have to
ascend slopes which follow each other in succession, each
being separated from its neighbour by a space of comparatively level ice. The slopes increase in steepness as we
ascend; they are steepest, moreover, on the right-hand
side of the glacier, where it is bounded by that from the
Périades, and at length we are unable to climb them without
the aid of an axe. Soon afterwards the dislocation of the

R 3

glacier becomes considerable ; we are lost in the clefts and depressions of the ice, and are unable to obtain a view sufficiently commanding to subdue these local appearances and convey to us the general aspect. We have at all events satisfied ourselves that on the upper portion of the glacier a succession of undulations exist, which sweep transversely across it. The term "wrinkles," applied to them by Prof. Forbes, is highly suggestive of the appearance which they present.

From the Cleft-station bands of snow may also be seen partially crossing the glacier in correspondence with the undulations upon its surface. If the quantity deposited the winter previous be large, and the heat of summer not too great, these bands extend quite across the glacier. They were first observed by Professor Forbes in 1843. In his Fifth Letter is given an illustrative diagram, which, though erroneous as regards the position of the veined structure, is quite correct in limiting the snow-bands to the Glacier du Géant proper.

At the place where the three welded tributaries of the Mer de Glace squeeze themselves through the strait of Tréla-porte, the bands undergo a considerable modification in shape. Near their origin they sweep across the Glacier du Géant in gentle curves, with their convexities directed downwards ; but at Trélaporte these curves, the chords of which a short time previous measured a thousand yards in length, have to squeeze themselves through a space of four hundred and ninety-five yards wide ; and as might be expected, they are here suddenly sharpened. The apex of each being thrust forward, they take the form of sharp hyperbolas, and preserve this character throughout the entire length of the Mer de Glace.

I would now conduct the reader to a point from which a good general view of the ice cascade of the Géant is attainable. From the old moraine near the lake of the

Tacul we observe the ice, as it descends the fall, to be broken into a succession of precipices. It would appear as if the glacier had its back periodically broken at the summit of the fall, and formed a series of vast chasms separated from each other by cliffy ridges of corresponding size. These, as they approach the bottom of the fall, become more and more toned down by the action of sun and air, and at some distance below the base of the cascade they are subdued so as to form the transverse undulations already described. These undulations are more and more reduced as the glacier descends; and long before the Tacul is attained, every sensible trace of them has disappeared. The terraces of the ice-fall are referred to by Professor Forbes in his Thirteenth Letter, where he thus describes them :—" The ice-falls succeed one another at regulated intervals, which appear to correspond to the renewal of each summer's activity in those realms of almost perpetual frost, when a swifter motion occasions a more rapid and wholesale projection of the mass over the steep, thus forming curvilinear terraces like vast stairs, which appear afterwards by consolidation to form the remarkable protuberant wrinkles on the surface of the Glacier du Géant."

With regard to the cause of the distribution of the dirt in bands, Professor Forbes writes thus in his Third Letter:— "I at length assured myself that it was entirely owing to the structure of the ice, which retains the dirt diffused by avalanches and the weather on those parts which are most porous, whilst the compacter portion is washed clean by the rain, so that those bands are nothing more than visible traces of the direction of the internal icy structure." Professor Forbes's theory, at that time, was that the glacier is composed throughout of a series of alternate segments of hard and porous ice, in the latter of which the dirt found a lodgment. I do not know whether he now retains his first opinion; but in his Fifteenth Letter he speaks of accounting for "the less compact structure of the ice beneath the dirt-band."

It appears to me that in the above explanation cause has been mistaken for effect. The ice on which the dirt-bands rest certainly appears to be of a spongier character than the cleaner intermediate ice; but instead of this being the cause of the dirt-bands, the latter, I imagine, by their more copious absorption of the sun's rays and the consequent greater disintegration of the ice, are the cause of the apparent porosity. I have not been able to detect any relative porosity in the "internal icy structure," nor am I able to find in the writings of Professor Forbes a description of the experiments whereby he satisfied himself that this assumed difference exists.

Several days of the summer of 1857 were devoted by me to the examination of these bands. I then found the bases and the frontal slopes of the undulations to which I have referred covered with a fine brown mud. These slopes were also, in some cases, covered with snow which the great heat of the weather had not been able entirely to remove. At places where the residue of snow was small its surface was exceedingly dirty—so dirty indeed that it appeared as if peat-mould had been strewn over it; its edges particularly were of a black brown. It was perfectly manifest that this snow formed a receptacle for the fine dirt transported by the innumerable little rills which trickled over the glacier. The snow gradually wasted, but it left its sediment behind, and thus each of the snowy bands observed by Professor Forbes in 1843, contributed to produce an appearance perfectly antithetical to its own. I have said that the frontal slopes of the undulations were thus covered; and it was on these, and not in the depressions, that the snow principally rested. The reason of this is to be found in the *bearing* of the Glacier du Géant, which, looking downwards, is about fourteen degrees east of the

meridian.* Hence the frontal slopes of the undulations
have a *northern aspect*, and it is this circumstance which,
in my opinion, causes the retention of the snow upon them.
Irrespective of the snow, the mere tendency of the dirt to
accumulate at the bases of the undulations would also
produce bands, and indeed does so on many glaciers ;
but the precision and beauty of the dirt-bands of the Mer
de Glace are, I think, to be mainly referred to the interception by the snow of the fine dark mud before referred
to on the northern slopes of its undulations.

Were the statements of some writers upon this subject
well founded, or were the dirt-bands as drawn upon the
map of Professor Forbes correctly shown, this explanation
could not stand a moment. It has been urged that the dirt-
bands cannot thus belong to a single tributary of the Mer
de Glace ; for if they did, they would be confined to that
tributary upon the trunk-glacier; whereas the fact is that
they extend quite across the trunk, and intersect the
moraines which divide the Glacier du Géant from its fel-
low-tributaries. From my first acquaintance with the
Mer de Glace I had reason to believe that this statement
was incorrect ; but last year I climbed a third time to the
Cleft-station for the purpose of once more inspecting the
bands from this fine position. I was accompanied by
Dr. Frankland and Auguste Balmat, and I drew the attention of both particularly to this point. Neither of them
could discern, nor could I, the slightest trace of a dirt-
band crossing any one of the moraines. Upon the trunk-
stream they were just as much confined to the Glacier
du Géant as ever. If the bands even existed east of the

* In the large map of Professor Forbes the bearing of the valley is
nearly sixty degrees west of the meridian; but this is caused by the true
north being drawn on the wrong side of the magnetic north; thus making
the declination easterly instead of westerly. In the map in Johnson's
' Physical Atlas ' this mistake is corrected.

moraines, they could not be seen, the dirt on this part of the glacier being sufficient to mask them.

The following interesting fact may perhaps have contributed to the production of the error referred to. Opposite to Trélaporte the eastern arms of the dirt-bands run so obliquely into the moraine of La Noire that the latter appears to be a tangent to them. But this moraine runs along the Mer de Glace, not far from its centre, and consequently the point of contact of each dirt-band with the moraine moves more quickly than the point of contact of the western arm of the same band with the side of the valley. Hence there is a tendency to *straighten* the bands; and at some distance down the glacier the effect of this is seen in the bands abutting against the moraine of La Noire at a larger angle than before. The branches thus abutting have, I believe, been ideally prolonged across the moraines.

On the map published by Prof. Forbes in 1843 the bands are shown crossing the medial moraines of the Mer de

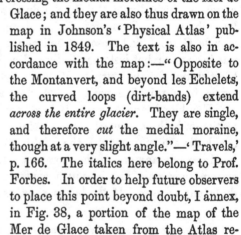

Fig. 38.

Glace; and they are also thus drawn on the map in Johnson's 'Physical Atlas' published in 1849. The text is also in accordance with the map:—"Opposite to the Montanvert, and beyond les Echelets, the curved loops (dirt-bands) extend *across the entire glacier*. They are single, and therefore *cut* the medial moraine, though at a very slight angle."—'Travels,' p. 166. The italics here belong to Prof. Forbes. In order to help future observers to place this point beyond doubt, I annex, in Fig. 38, a portion of the map of the Mer de Glace taken from the Atlas referred to. If it be compared with Fig. 35 the difference between Prof. Forbes and myself will be clearly seen. The portion of the glacier represented in both diagrams may be viewed from the point near the Flegère already referred to.

The explanation which I have given involves three considerations:—The transverse breaking of the glacier on the cascade, and the gradual accumulation of the dirt in the hollows between the ridges; the subsequent toning down of the ridges to gentle protuberances which sweep across the glacier; and the collection of the dirt upon the slopes and at the bases of these protuberances. Whether the periods of transverse fracture are annual or not—whether the "wrinkles" correspond to a yearly gush—and whether, consequently, the dirt-bands mark the growth of a glacier as the "annual rings" mark the growth of a tree, I do not know. It is a conjecture well worthy of consideration; but it is only a conjecture, which future observation may either ratify or refute.

THE VEINED STRUCTURE OF GLACIERS.

(27.)

THE general appearance of the veined structure may be thus briefly described :—The ice of glaciers, especially midway between their mountain-sources and their inferior extremities, is of a whitish hue, caused by the number of small air-bubbles which it contains, and which, no doubt, constitute the residue of the air originally entrapped in the interstices of the snow from which it has been derived. Through the general whitish mass, at some places, innumerable parallel veins of clearer ice are drawn, which usually present a beautiful blue colour, and give the ice a laminated appearance. The cause of the blueness is, that the air-bubbles, distributed so plentifully through the general mass, do not exist in the veins, or only in comparatively small numbers.

In different glaciers, and in different parts of the same glacier, these veins display various degrees of perfection. On the clean unweathered walls of some crevasses, and in the channels worn in the ice by glacier-streams, they are most distinctly seen, and are often exquisitely beautiful. They are not to be regarded as a partial phenomenon, or as affecting the constitution of glaciers to a small extent merely. A large portion of the ice of some glaciers is thus affected. The greater part, for example, of the Mer de Glace consists of this laminated ice ; and the whole of the Glacier of the Rhone, from the base of the ice-cascade downwards, is composed of ice of the same description.

Those who have ascended Snowdon, or wandered among the hills of Cumberland, or even walked in the environs of Leeds, Blackburn, and other towns in Yorkshire and Lancashire, where the stratified sandstone of the district is used for building purposes, may have observed the weathered edges of the slate rocks or of the building-stone to be grooved and furrowed. Some laminæ of such rocks withstand the action of the atmosphere better than others, and the more resistent ones stand out in ridges after the softer parts between them have been eaten away. An effect exactly similar is observed where the laminated ice of glaciers is exposed to the action of the sun and air. Little grooves and ridges are formed upon its surface, the more resistent plates protruding after the softer material between them has been melted away.

One consequence of this furrowing is, that the light dirt scattered by the winds over the surface of the glacier is gradually washed into the little grooves, thus forming fine lines resembling those produced by the passage of a rake over a sanded walk. These lines are a valuable index to some of the phenomena of motion. From a position on the ice of the Glacier du Géant a little higher up than Tré-laporte a fine view of these superficial groovings is obtained ; but the dirt-lines are not always straight. A slight power of independent motion is enjoyed by the separate parts into which a glacier is divided by its crevasses and dislocations, and hence it is, that, at the place alluded to, the dirt-lines are bent hither and thither, though the ruptures of conti-nuity are too small to affect materially the general direction of the structure. On the glacier of the Talèfre I found these groovings useful as indicating the character of the forces to which the ice near the summit of the fall is sub-jected. The ridges between the chasms are in many cases violently bent and twisted, while the adjacent groovings enable us to see the normal position of the mass.

The veined structure has been observed by different travellers; but it was probably first referred to by Sir David Brewster, who noticed the veins of the Mer de Glace on the 10th of September, 1814. It was also

observed by General Sabine,* by Rendu, by Agassiz, and
no doubt by many others; but the first clear description
of it was given by M. Guyot, in a communication presented to the Geological Society of France in 1838. I quote
the following passage from this paper:—"I saw under my
feet the surface of the entire glacier covered with regular
furrows from one to two inches wide, hollowed out in a half
snowy mass, and separated by protruding plates of harder
and more transparent ice. It was evident that the mass
of the glacier here was composed of two sorts of ice, one
that of the furrows, snowy and more easily melted; the
other that of the plates, more perfect, crystalline, glassy,
and resistent; and that the unequal resistance which the
two kinds of ice presented to the atmosphere was the cause
of the furrows and ridges. After having followed them for
several hundreds of yards, I reached a fissure twenty or thirty
feet wide, which, as it cut the plates and furrows at right
angles, exposed the interior of the glacier to a depth of

* In reply to a question in connexion with this subject, General Sabine
has favoured me with the following note:—

"MY DEAR TYNDALL,

"It was in the summer of 1841, at the Lower Grindelwald Glacier,
that I first saw, and was greatly impressed and interested by examining
and endeavouring to understand (in which I did not succeed), the veined
structure of the ice. I do not remember when I mentioned it to Forbes,
but it must be before 1843, because it is noticed in his book, p. 29. I had
never observed it in the glaciers of Spitzbergen or Baffin's Bay, or in the
icebergs of the shores and straits of Davis or Barrow. I feel the more
confident of this, because, when I first saw the veined structure in Switzerland, my arctic experience was more fresh in my recollection, and I recol-
lected nothing like it.

"Veins are indeed not uncommon in icebergs, but they quite resemble
veins in rocks, and are formed by water filling fissures and freezing into
blue ice, finely contrasted with the white granular substance of the berg.

"The ice of the Grindelwald Glacier (where I examined the veined
structure) was broken up into very large masses, which by pressure had
been upturned, so that a very poor judgment would be formed of the direction of the veins as they existed in the glacier before it had broken up.

"Sincerely yours,

"Feb. 20, 1860."   "EDWARD SABINE.

thirty or forty feet, and gave a beautiful transverse section
of the structure. As far as my vision could reach I saw
the mass of the glacier composed of layers of snowy ice,
each two of which were separated by one of the plates of
which I have spoken, the whole forming a regularly laminated mass, which resembled certain calcareous slates."

Previous observers had mistaken the lamination for
stratification; but M. Guyot not only clearly saw that they
were different, but in the comparison which he makes he
touches, I believe, on the true cause of the glacier-structure.
He did not hazard an explanation of the phenomenon, and
I believe his memoir remained unprinted. In 1841 the
structure was noticed by Professor Forbes during his visit
to M. Agassiz on the Lower Aar Glacier, and described in
a communication presented by him to the Royal Society
of Edinburgh. He subsequently devoted much time to
the subject, and his great merit in connexion with it
consists in the significance which he ascribed to the phenomenon when he first observed it, and in the fact of his
having proved it to be a constitutional feature of glaciers
in general.

The first explanation given of those veins by Professor
Forbes was, that they were small fissures formed in the ice
by its motion; that these were filled with the water of the
melted ice in summer, which froze in winter so as to form
the blue veins. This is the explanation given in his
'Travels,' page 377; and in a letter published in the
'Edinburgh New Philosophical Journal,' October, 1844, it is
re-affirmed in these words:—"With the abundance of blue
bands before us in the direction in which the differential
motion must take place (in this case sensibly parallel to the
sides of the glacier), it is impossible to doubt that these
infiltrated crevices (for such they undoubtedly are) have
this origin." This theory was examined by Mr. Huxley
and myself in our joint paper; but it has been since alleged

that ours was unnecessary labour, Prof. Forbes himself
having in his Thirteenth Letter renounced the theory,
and substituted another in its place. The latter theory
differs, so far as I can understand it, from the former in
this particular, that the *freezing of the water* in the fissures
is discarded, their sides being now supposed to be united
"by the simple effects of time and cohesion."*   For a
statement of the change which his opinions have undergone,
I would refer to the Prefatory Note which precedes the
volume of 'Occasional Papers' recently published by Prof.
Forbes; but it would have diminished my difficulty had
the author given, in connexion with his new volume, a
more distinct statement of his present views regarding
the veined structure. With many of his observations
and remarks I should agree; with many others I cannot
say whether I agree or not; and there are others still
with which I do not think I should agree: but in hardly
any case am I certain of his precise views, excepting,
indeed, the cardinal one, wherein he and others agree in
ascribing to the structure a different origin from stratification.   Thus circumstanced, my proper course, I think,
will be to state what I believe to be the cause of the
structure, and leave it to the reader to decide how far
our views harmonize; or to what extent either of them is
a true interpretation of nature.

Most of the earlier observers considered the structure to
be due to the stratification of the mountain-snows—a view
which has received later development at the hands of Mr.
John Ball; and the practical difficulty of distinguishing
the undoubted effects of *stratification* from the phenomena

---

* In a letter to myself, published in the 17th volume of the 'Philosophical Magazine,' Professor Forbes writes as follows :—" In 1846, then, I
abandoned no part of the theory of the veined structure, on which as you
say so much labour had been expended, except the admission, always
yielded with reluctance, and got rid of with satisfaction, that the conge-
lation of water in the crevices of the glacier may extend in winter to a
great depth."

presented by *structure*, entitles this view to the fullest
consideration.   The blue veins of glaciers are, however,
not always, nor even generally, such as we should expect
to result from stratification.   The latter would furnish us
with distinct planes extending parallel to each other for
considerable distances through the glacier ; but this,
though sometimes the case, is by no means the general
character of the structure.   We observe blue streaks, from
a few inches to several feet in length, upon the walls of
the same crevasse, and varying from the fraction of an
inch to several inches in thickness.   In some cases the
streaks are definitely bounded, giving rise to an appearance resembling the section of a lens, and hence called
the "lenticular structure" by Mr. Huxley and myself; but
more usually they fade away in pale washy streaks through
the general mass of the whitish ice.   In Fig. 39 I have
given a representation of the
structure as it is very commonly exhibited on the walls
of crevasses. Its aspect is not
that which we should expect
from the consolidation of successive beds of mountain snow.

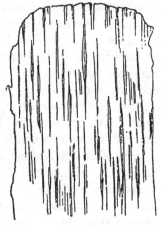

Fig. 39.

   Further, at the bases of ice-
cascades the structural laminæ are usually *vertical :* below
the cascade of the Talèfre,
of the Noire, of the Strahleck branch of the Lower
Grindelwald Glacier, of the
Rhone, and other ice-falls,
this is the case; and it seems extremely difficult to
conceive that a mass horizontally stratified at the summit
of the fall, should, in its descent, contrive to turn its strata
perfectly on end.

Again, we often find a very feebly-developed structure at the central portions of a glacier, while the lateral portions are very decidedly laminated. This is the case where the inclination of the glacier is nearly uniform throughout; and where no medial moraines occur to complicate the phenomenon. But if the veins mark the bedding, there seems to be no sufficient reason for their appearance at the lateral portions of the glacier, and their absence from the centre.

This leads me to the point at which what I consider to be the true cause of the structure may be referred to. The theoretic researches of Mr. Hopkins have taught us a good deal regarding the pressures and tensions consequent upon glacier-motion. Aided by this knowledge, and also by a mode of experiment first introduced by Professor Forbes, I will now endeavour to explain the significance of the fact referred to in the last paragraph. If a plastic substance, such as mud, flow down a sloping canal, the lateral portions, being held back by friction, will be outstripped by the central ones. When the flow is so regulated that the velocity of a point at the centre shall not vary throughout the entire length of the canal, a coloured circle stamped upon the centre of the mud stream, near its origin, will move along with the mud, and still retain its circular form; for, inasmuch as the velocity of all points along the centre is the same, there can be no elongation of the circle longitudinally or transversely by either strain or pressure. A similar absence of longitudinal pressure may exist in a glacier, and, where it exists throughout, no central structure can, in my opinion, be developed.

But let a circle be stamped upon the mud-stream near its side, then, when the mud flows, this circle will be distorted to an oval, with its major axis oblique to the direction of motion; the cause of this is that the portion of the circle farthest from the side of the canal moves more freely than that adjacent to the side. The mechanical effect of the slower lateral motion is to squeeze the circle in one direction, and draw it out in the perpendicular one.

A glance at Fig. 40 will render all that I have said in-telligible. The three cir-cles are first stamped on the mud in the same transverse line; but after they have moved downwards they will be in the same straight line no longer. The central one will be the foremost; while the lateral ones have their forms changed from circles to ovals. In a glacier of the shape of this canal exactly similar effects are produced. Now the shorter axis *m n* of each oval is a line of squeezing or pressure; the longer axis is a line of strain or tension; and the associated glacier-phenomena are as follows:—Across the line *m n*, or perpendicular to the pressure, we have the *veined structure* developed, while across the line of tension the glacier usually breaks and forms *marginal crevasses.* Mr. Hopkins has shown that the lines of greatest pressure and of greatest strain are at right angles to each other, and that in valleys of a uniform width they enclose an angle of forty-five degrees with the side of the glacier. To the structure thus formed I have applied the term *marginal structure.* Here, then, we see that there are mechanical agencies at work near the side of such a glacier which are absent from the centre, and we have effects developed—I believe *by the pressure*—in the lateral ice, which are not produced in the central.

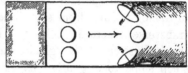

Fig. 40.

I have used the term "uniform inclination" in connexion with the marginal structure, and my reason for doing so will now appear. In many glaciers the structure,

instead of being confined to the margins, sweeps quite
across them.   This is the case, for example, on the Glacier
du Géant, the structure of which is prolonged into the
Mer de Glace.   In passing the strait at Trélaporte, however, the curves are squeezed and their apices bruised,
so that the structure is thrown into a state of confusion;
and thus upon the Mer de Glace we encounter difficulty
in tracing it fairly from side to side.   Now the key to this
transverse structure I believe to be the following: Where
the inclination of the glacier suddenly changes from a
steep slope to a gentler, as at the bases of the "cascades,"—the ice to a certain depth must be thrown into
a state of violent longitudinal compression; and along
with this we have the resistance which the gentler slope
throws athwart the ice descending from the steep one.
At such places a structure is developed transverse to
the axis of the glacier, and likewise transverse to the
pressure.   The quicker flow of the centre causes this structure to bend more and more, and after a time it sweeps in
vast curves across the entire glacier.

  In illustration of this point I will refer, in the first place,
to that tributary of the Lower Glacier of Grindelwald
which descends from the Strahleck.   Walking up this tributary we come at length to the base of an ice-fall.   Let
the observer here leave the ice, and betake himself to
either side of the flanking mountain.   On attaining a point
which commands a view both of the fall and of the glacier
below it, an inspection of the glacier will, I imagine, solve
to his satisfaction the case of structure now under consideration.

  It is indeed a grand experiment which Nature here
submits to our inspection.   The glacier descending from
its *névé* reaches the summit of the cascade, and is broken
transversely as it crosses the brow; it afterwards descends
the fall in a succession of cliffy ice-ridges with transverse

hollows between them. In these latter the broken ice and débris collect, thus partially choking the fissures formed in the first instance. Carrying the eye downwards along the fall, we see, as we approach the base, these sharp ridges toned down; and a little below the base they dwindle into rounded protuberances which sweep in curves quite across the glacier. At the base of the fall the structure begins to appear, feebly at first, but becoming gradually more pronounced, until, at a short distance below the base of the fall, the eye can follow the fine superficial groovings from side to side; while at the same time the ice underneath the surface has become laminated in the most beautiful manner.

It is difficult to convey by writing the force of the evidence which the actual observation of this natural experiment places before the mind. The ice at the base of the fall, retarded by the gentler inclination of the valley, has to bear the thrust of the descending mass, the sudden change of inclination producing powerful longitudinal compression. The protuberances are squeezed more closely together, the hollows between them appear to wrinkle up in submission to the pressure—in short, the entire aspect of the glacier suggests the powerful operations of the latter force. At the place where *it* is exerted the veined structure makes its appearance; and being once formed, it moves downwards, and gives a character to other portions of the glacier which had no share in its formation.

An illustration almost as good, and equally accessible, is furnished by the Glacier of the Rhone. I have examined the grand cascade of this glacier from both sides; and an ordinary mountaineer will find little difficulty in reaching a point from which the fall and the terminal portion of the glacier are b th distinctly visible. Here also he will find the cliffy ridges separated from

S

each other by transverse chasms, becoming more and more subdued at the bottom of the fall, and disappearing entirely lower down the glacier. As in the case of the Grindelwald Glacier, the squeezing of the protuberances and of the spaces between them is quite apparent, and where this squeezing commences the transverse structure makes its appearance. All the ice that forms the lower portion of this glacier has to pass through the *structure-mill* at the bottom of the fall, and the consequence is that *it is all laminated.*

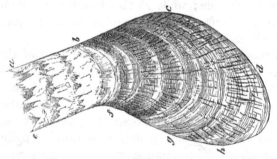

Fig. 41.

This case of structural development will be better appreciated on reference to Figs. 41 and 42, the former of which

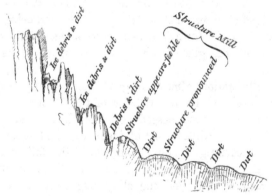

Fig. 42.

is a plan, and the latter a section, of a part of the
ice-fall and of the glacier below it; *a b c f* is the gorge
of the fall, *f b* being the base. The transverse cliffy
ice-ridges are shown crossing the cascade, being subdued
at the base to protuberances which gradually disappear as
they advance downwards. The structure sweeps over the
glacier in the direction of the fine curved lines; and I have
also endeavoured to show the direction of the radial crevasses, which, in the centre at least, are at right angles to
the veins. To the manifestation of structure here considered I have, for the sake of convenient reference, applied
the term *transverse structure*.

A third exhibition of the structure is now to be
noticed. We sometimes find it in the *middle* of a glacier
and running *parallel* to its length. On the centre of the
ice-fall of the Talèfre, for example, we have a structure of
this kind which preserves itself parallel to the axis of the
fall from top to bottom. But we discover its origin higher
up. The structure here has been produced at the extremity
of the Jardin, where the divided ice meets, and not only
brings into partial parallelism the veins previously existing along the sides of the Jardin, but develops them still
further by the mutual pressure of the portions of newly
welded ice. Where two tributary glaciers unite, this is
perhaps without exception the case. Underneath the
moraine formed by the junction of the Talèfre and Léchaud
the structure is finely developed, and the veins run in the
direction of the moraine. The same is true of the ice
under the moraine formed by the junction of the Léchaud
and Géant. These afterwards form the great medial moraines of the Mer de Glace, and hence the structure of the
trunk-stream underneath these moraines is parallel to the
direction of the glacier. This is also true of the system of
moraines formed by the glaciers of Monte Rosa. It is
true in an especial manner of the lower glacier of the Aar,

whose medial moraine perhaps attains grander proportions than any other in the Alps, and underneath which the structure is finely developed.

The manner in which I have illustrated the production of this structure will be understood from Fig. 43. B B′ are two wooden boxes, communicating by sluice-fronts with two branch canals, which unite to a common trunk at G. They are intended to represent respectively the trunk and tributaries of the Unteraar Glacier, the part G being the Abschwung, where the Lauteraar and Finsteraar glaciers unite to form the Unteraar. The mud is first permitted to flow beneath the two sluices until it has covered the bottom of the trough for some distance, when it is arrested. The end of a glass tube is then dipped into a mixture of rouge and water, and small circles are stamped upon the mud. The two branches are thickly covered with these circles. The sluices being again raised, the mud in the branches moves downwards, carrying with it the circles stamped upon it; and the manner in which these circles are distorted enables us to infer the strains and pressures to which the mud is subjected during its descent. The figure represents approximately what takes place. The side-circles, as might be expected, are squeezed to oblique ovals, but it is at the junction of the branches that the chief effect of pressure is

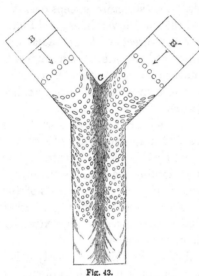

Fig. 43.

produced. Here, by the mutual thrust of the branches, the circles are not only changed to elongated ellipses, but even squeezed to straight lines. In the case of the glacier this is the region at which the structure receives its main development. To this manifestation of the veins I have applied the term *longitudinal structure.*

The three main sources of the blue veins are, I think, here noted ; but besides these there are many local causes which influence their production. I have seen them well formed where a glacier is opposed by the sudden bend of a valley, or by a local promontory which presents an obstacle sufficient to bring the requisite pressure into play. In the glaciers of the Tyrol and of the Oberland I have seen examples of this kind; but the three principal sources of the veins are, I think, those stated above.

It was long before I cleared my mind of doubt regarding the origin of the lamination. When on the Mer de Glace in 1857 I spared neither risk nor labour to instruct myself regarding it. I explored the Talèfre basin, its cascade, and the ice beneath it. Several days were spent amid the ice humps and cliffs at the lower portion of the fall. I suppose I traversed the Glacier du Géant twenty times, and passed eight or ten days amid the confusion of its great cascade. I visited those places where, it had been affirmed, the veins were produced. I endeavoured to satisfy myself of the mutability which had been ascribed to them ; but a close examination reduced the value of each particular case so much that I quitted the glacier that year with nothing more than an *opinion* that the structure and the stratification were two different things. I, however, drew up a statement of the facts observed, with the view of presenting it to the Royal Society ; but I afterwards felt that in thus acting I should merely swell the literature of the subject without adding anything certain. I therefore withheld the paper, and

resolved to devote another year to a search among the chief
glaciers of the Oberland, of the Canton Valais, and of Savoy,
for proofs which should relieve my mind of all doubt upon
the subject.

Accordingly in 1858 I visited the glaciers of Rosenlaui,
Schwartzwald, Grindelwald, the Aar, the Rhone, and the
Aletsch, to the examination of which latter I devoted more
than a week. I afterwards went to Zermatt, and, taking up
my quarters at the Riffelberg, devoted eleven days to the
examination of the great system of glaciers of Monte Rosa.
I explored the Görner glacier up almost to the Cima de
Jazzi; and believed that in it I could trace the structure
from portions of the glacier where it vanished, through
various stages of perfection, up to its full development. I
believe this still; but yet it is nothing but a belief, which
the utmost labour that I could bestow did not raise
to a certainty. The Western glacier of Monte Rosa, the
Schwartze glacier, the Trifti glacier, the glacier of the
little Mont Cervin, and of St. Théodule, were all examined
in connexion with the great trunk-stream of the Görner.
to which they weld themselves; and though the more I
pursued the subject the stronger my conviction became
that pressure was the cause of the structure, a crucial case
was still wanting.

In the phenomena of slaty cleavage, it is often, if not
usually, found that the true cleavage *cuts* the planes of
stratification—sometimes at a very high angle. Had this
not been proved by the observations of Sedgwick and
others, geologists would not have been able to conclude that
cleavage and bedding were two different things, and
needed wholly different explanations. My aim, throughout
the expedition of 1858, was to discover in the ice a parallel
case to the above; to find a clear and undoubted instance
where the veins and the stratification were simultaneously
exhibited, cutting each other at an unmistakeable angle.

On the 6th of August, while engaged with Professor Ramsay upon the Great Aletsch Glacier, not far from its junction with the Middle Aletsch, I observed what appeared to me to be the lines of bedding running nearly horizontal along the wall of a great crevasse, while cutting them at a large angle was the true veined structure. I drew my friend's attention to the fact, and to him it appeared perfectly conclusive. It is from a sketch made by him at the place that Fig. 44 has been taken.

Fig. 44.

This was the only case of the kind which I observed upon the Aletsch Glacier; and as I afterwards spent day after day upon the Monte Rosa glaciers, vainly seeking a similar instance, the thought again haunted me that we might have been mistaken upon the Aletsch. In this state of mind I remained until the 18th of August, a day devoted to the examination of the Furgge Glacier, which lies at the base of the Mont Cervin.

Crossing the valley of the Görner Glacier, and taking a plunge as I passed into the Schwarze See, I reached, in good time, the object of my day's excursion. Walking up the glacier, I at length found myself opposed by a frozen cascade composed of four high terraces of ice. The highest of these was chiefly composed of ice-cliffs and *séracs*, many of which had fallen, and now stood like rocking-stones upon the edge of the second terrace. The glacier at the base of the cascade was strewn with broken ice, and some blocks two hundred cubic feet in volume had been cast to a considerable distance down the glacier.

Upon the faces of the terraces the stratification of the *névé* was most beautifully shown, running in parallel and horizontal lines along the weathered surface. The snow-field above the cascade is a frozen plain, smooth almost as a sheltered lake. The successive snow-falls deposit themselves with great regularity, and at the summit of the cascade the sections of the *névé* are for the first time exposed. Hence their peculiar beauty and definition.

Indeed the figure of a lake pouring itself over a rocky barrier which curves convexly upwards, thus causing the water to fall down it, not only longitudinally over the vertex of the curve, but laterally over its two arms, will convey a tolerably correct conception of the shape of the fall. Towards the centre the ice was powerfully squeezed laterally, the beds were bent, and their continuity often broken by faults. On inspecting the ice from a distance with my opera glass, I thought I saw structural groovings cutting the strata at almost a right angle. Had the question been an undisputed one, I should perhaps have felt so sure of this as not to incur the danger of pushing the inquiry further; but, under the circumstances, danger was a secondary point. Resigning, therefore, my glass to my guide, who was to watch the tottering blocks overhead, and give me warning should they move, I advanced to the base of the fall, removed with my hatchet the weathered surface of the ice, and found underneath it the true veined structure, cutting, at nearly a right angle, the planes of stratification. The superficial groovings were not uniformly distributed over the fall, but appeared most decided at those places where the ice appeared to have been most squeezed. I examined three or four of these places, and in each case found the true veins nearly vertical, while the bedding was horizontal. Having perfectly satisfied myself of these facts, I made a speedy retreat, for the ice-blocks seemed most threatening, and the sunny hour was that at which they fall most frequently.

I next tried the ascent of the glacier up a dislocated declivity to the right. The ice was much riven, but still practicable. My way for a time lay amid fissures which exposed magnificent sections, and every step I took added further demonstration to what I had observed below. The strata were perfectly distinct, the structure equally so, and one crossed the other at an angle of seventy or eighty degrees. Mr. Sorby has adduced a case of the crumpling of a bed of sandstone through which the cleavage passes: here on the glacier I had parallel cases; the beds were bent and crumpled, but the structure ran through the ice in sharp straight lines. This perhaps was the most pleasant day I ever spent upon the glaciers: my mind was relieved of a long brooding doubt, and the intellectual freedom thus obtained added a subjective grandeur to the noble scene before me. Climbing the cliffs near the base of the Matterhorn, I walked along the rocky spine which extends to the Hörnli, and afterwards descended by the valley of Zmutt to Zermatt.

A year after my return to England a remark contained in Professor Mousson's interesting little work 'Die Gletscher der Jetzzeit' caused me to refer to the atlas of M. Agassiz's 'Système Glaciaire,' from which I learned that this indefatigable observer had figured a case of stratification and structure cutting each other. If, however, I had seen this figure beforehand, it would not have changed my movements; for the case, as sketched, would not have convinced me. I have now no doubt that M. Agassiz has

s 3

preceded me in this observation, and hence my results are
to be taken as mere confirmations of his.

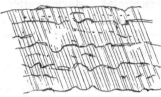

Fig. 45 represents a
crumpled portion of the ice
with the lines of lamination
passing through the strata.
Fig. 46 represents a case
where a fault had oc-
curred, the veins at both

Fig. 45.

sides of the line of dislocation being inclined towards each
other.

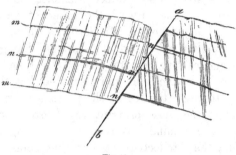

Fig. 46.

THE VEINED STRUCTURE AND THE DIFFERENTIAL MOTION.

(28.)

I HAVE now to examine briefly the explanation of the structure which refers it to differential motion—to a sliding of the particles of ice past each other, which leaves the traces of its existence in the blue veins. The fact is emphatically dwelt upon by those who hold this view, that the structure is best developed nearest to the sides of the glacier, where the differential motion is greatest. Why the differential motion is at its maximum near to the sides is easily understood. Let A B, C D, Fig. 47, represent the two sides of a glacier, moving in the direction of the arrow, and let $m\ a\ b\ c\ n$ be a straight line of stakes set out across the glacier to-day. Six months hence this line, by the motion of the ice downwards, will be bent to the form $m\ a'\ b'\ c'\ n$: this curve will not be circular, it will be flattened in the middle ; the points a and c, at some distance on each side of the centre b, move in

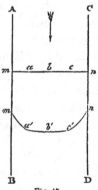

Fig. 47.

fact with nearly the same velocity as the centre itself. Not so with the sides:—a' and c' have moved considerably in advance of m and n, and hence we say that the difference of motion, or the differential motion, of the particles of ice near to the side is a maximum.

During all this time the points $m\ a'\ b'\ c'\ n$ have been moving straight down the glacier; and hence it will be understood that the sliding of the parts past each other, or, in other words, the differential motion, *is parallel to the sides of the glacier*. This, indeed, is the only differential motion that experiment has ever established; and consequently, when we find the best blue veins referred to the sides of the glacier because the differential motion is there greatest, we naturally infer that the motion meant is parallel to the sides.

But the fact is, that this motion would not at all account for the blue veins, for they are not parallel to the sides, but *oblique* to them. This difficulty revealed itself after a time to those who first propounded the theory of differential motion, and caused them to modify their explanation of the structure. Differential motion is still assumed to be the cause of the veins, but now a motion is meant oblique to the sides, and it is supposed to be obtained in the following way:—Through the quicker motion of the point c' the ice between it and n becomes distended; that is to say, the line $c'\, n$ is in a state of strain—there is a *drag*, it is said, oblique to the sides of the glacier; and it is therefore in this direction that the particles will be caused to slide past each other. Dr. Whewell, who advocates this view, thus expounds it. He supposes the case of an alpine valley filled with India-rubber which has been warmed until it has partially melted, or become viscous, and then asks, " What will now be the condition of the mass? The sides and bottom will still be held back by the friction; the middle and upper part will slide forwards, but not freely. This want of freedom in the motion (arising from the viscosity) will produce a drag towards the middle of the valley, where the motion is freest; hence the direction in which the filaments slide past each other will be obliquely directed towards the middle. The sliding will separate the mass according to such lines; and though new attachments will take place, the mass may be expected to retain the results of this separation in the traces of parallel fissures."*   Nothing can be clearer than the image of the process thus placed before the mind's eye.

One fact of especial importance is to be borne in mind : the sliding of filaments which is thus supposed to take place oblique to the glacier has never been proved; it is wholly assumed.   A moraine, it is admitted, will run parallel to the side of a glacier, or a block will move in the same direction from beginning to end, without being sensibly drawn towards the centre, but still it is supposed that the sliding of parts exists, though of a character so small as to render it insensible to measurement.

My chief difficulty as regards this theory may be expressed in a very few words.   If the structure be produced by differential motion, why is the large and *real* differential motion which experiments have established incompetent to produce it?   And how can the veins run, as they are admitted to do, *across the lines of maximum sliding* from their origin throughout the glacier to its end?

That a drag towards the centre of the glacier exists is undeniable, but that in consequence of the drag there is a sliding of filaments in this direction, is quite another thing. I have in another place † endeavoured to show experimentally that no such sliding takes place, that the drag on any point towards the centre expresses only half the conditions of the problem; being exactly neutralized by the thrust towards the sides.   It has been, moreover, shown by Mr. Hopkins that the lines of maximum strain and of maximum sliding cannot coincide; indeed, if all the particles be urged by the same force, no matter how strong the pull may be, there will be no tendency of one to slide past the other.

* 'Philosophical Magazine,' Ser. III., vol. xxvi.
† 'Proceedings of the Royal Institution,' vol. ii. p. 324.

THE RIPPLE-THEORY OF THE VEINED STRUCTURE.

(29.)

THE assumption of oblique sliding, and the production thereby of the marginal structure, have, however, been fortified by considerations of an ingenious and very interesting kind. "How," I have asked, "can the oblique structure persist across the lines of greatest differential motion throughout the length of the glacier?" But here I am met by another question which at first sight might seem equally unanswerable—"How do ripple-marks on the surface of a flowing river, which are nothing else than lines of differential motion of a low order, cross the river from the sides obliquely, while the direction of greatest differential motion is parallel to the sides?" If I understand aright, this is the main argument of Professor Forbes in favour of his theory of the oblique marginal structure. It is first introduced in a note at page 378 of his 'Travels;' he alludes to it in a letter written the following year; in his paper in the 'Philosophical Transactions' he develops the theory. He there gives drawings of ripple-marks observed in smooth gutters after rain, and which he finds to be inclined to the course of the stream, exactly as the marginal structure is inclined to the side of the glacier. The explanation also embraces the case of an obstacle placed in the centre of a river. "A case," writes Professor Forbes, "parallel to the last mentioned, where a fixed obstacle cleaves a descending stream, and leaves its trace in the fan-shaped tail, is well known in several glaciers, as in that at Ferpêcle, and the Glacier de Lys on the south side of Monte Rosa; particularly the last, where the veined structure follows the law just mentioned." In his Twelfth Letter he also refers to the ripples " as exactly corresponding to the position of the icy bands." In his letter to Dr. Whewell, published in the 'Occasional Papers,' page 58, he writes as follows :—"The same is remarkably shown in the case of a stream of water, for instance a mill-race. Although the movement of the water, as shown by floating bodies, is exceedingly nearly (for small velocities sensibly) parallel to the sides, yet the variation of the speed from the side to the centre of the stream occasions a *ripple*, or molecular discontinuity, which inclines forwards from the sides to the centre of the stream at an angle with the axis depending on the ratio of the central and lateral velocity. The veined structure of the ice corresponds to the ripple of the water, a molecular discontinuity whose measure is not comparable to the actual velocity of the ice; and therefore the general movement of the glacier, as indicated by the moraines, remains sensibly parallel to the sides." This theory opens up to us a series of interesting and novel considerations which I think will repay the reader's attention. If the ripples in the water and the veins in the ice be due to the same mechanical cause, when we develop clearly the origin of the former we are led directly to the explanation of the latter. I shall now endeavour to reduce the ripples to their mechanical elements.

The Messrs. Weber have described in their 'Wellenlehre' an effect of wave-motion which it is very easy to obtain. When a boat moves through perfectly smooth water, and the rower raises his oar out of the water, drops trickle from its blade, and each drop where it falls produces a system of concentric rings. The circular waves as they widen become depressed, and, if the drops succeed each other with sufficient speed, the rings cross each other at innumerable points. The effect of this is to blot out more or less completely all the

circles, and to leave behind two straight divergent
ripple-lines, which are tangents to all the external rings;
being in fact formed by the intersections of the latter,
as a caustic in optics is formed by the intersection of
luminous rays. Fig. 48, which is virtually copied from

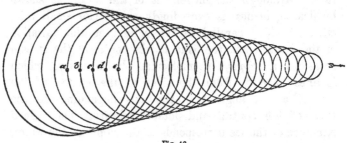

Fig. 48.

M. Weber, will render this description at once intelligible. The boat is supposed to move in the direction of
the arrow, and as it does so the rings which it leaves
behind widen, and produce the divergence of the two
straight resultant lines of ripple.

The more quickly the drops succeed each other, the
more frequent will be the intersections of the rings; but as
the speed of succession augments we approach the case
of *a continuous vein* of liquid; and if we suppose the continuity to be perfectly established, the ripples will still be
produced with a smooth space between them as before.
This experiment may indeed be made with a well-wetted
oar, which on its first emergence from the water sends into
it a continuous liquid vein. The same effect is produced
when we substitute for the stream of liquid a solid rod—
a common walking-stick for example. A water-fowl swimming in calm water produces two divergent lines of ripples
of a similar kind.

We have here supposed the water of the lake to be at
rest, and the liquid vein or the solid rod to move through

it; but precisely the same effect is produced if we suppose the rod at rest and the liquid in motion.   Let a post, for example, be fixed in the middle of a flowing river; diverging from that post right and left we shall have lines of ripples exactly as if the liquid were at rest and the post moved through it with the velocity of the river.   If the same post be placed close to the bank, so that *one* of its edges only shall act upon the water, diverging from that edge we shall have a *single* line of ripples which will cross the river obliquely towards its centre.   It is manifest that any other obstacle will produce the same effect as our hypothetical post.   In the words of Professor Forbes, "the slightest prominence of any kind in the wall of such a conduit, a bit of wood or a tuft of grass, is sufficient to produce a well-marked ripple-streak from the side towards the centre."

The foregoing considerations show that the divergence of the two lines of ripples from the central post, and of the single line in the case of the lateral post, have their mechanical element, if I may use the term, in the experiment of the Messrs. Weber.   In the case of a swimming duck the connexion between the diverging lines of ripples and the propagation of rings round a disturbed point is often very prettily shown.   When the creature swims with vigour the little foot with which it strikes the water often comes sufficiently near to the surface to produce an elevation,—sometimes indeed emerging from the water altogether.   Round the point thus disturbed rings are immediately propagated, and the widening of those rings *is the exact measure of the divergence of the ripple lines.*   The rings never cross the lines;—the lines never retreat from the rings.

If we compare the mechanical actions here traced out with those which take place upon a glacier, I think it will be seen that the analogy between the ripples and the

veined structure is entirely superficial. How the structure ascribed to the Glacier de Lys is to be explained I
do not know, for I have never seen it; but it seems impossible that it could be produced, as ripples are, by a fixed
obstacle which "cleaves a descending stream." No one
surely will affirm that glacier-ice so closely resembles a
fluid as to be capable of transmitting undulations, as water
propagates rings round a disturbed point. The difficulty
of such a supposition would be augmented by taking into
account the motion of the *individual liquid particles* which
go to form a ripple; for the Messrs. Weber have shown
that these move in closed curves, describing orbits more
or less circular. Can it be supposed that the particles of
ice execute a motion of this kind? If so, their orbital
motions may be easily calculated, being deducible from
the motion of the glacier compounded with the inclination
of the veins. If so important a result could be established,
all glacier theories would vanish in comparison with it.

There is another interesting point involved in the passage above quoted. Professor Forbes considers that the
ripple is occasioned by the variation of speed from the side
to the centre of the stream, and that its *inclination* depends
on the ratio of the central and lateral velocity. If I am
correct in the above analysis, this cannot be the case. The
inclination of the ripple depends solely on the ratio of the
river's translatory motion to the velocity of its wave-motion. Were the lateral and central velocities alike, a
momentary disturbance at the side would produce a *straight*
ripple-mark, whose inclination would be compounded of
the two elements just mentioned. If the motion of the
water vary from side to centre, the velocity of wave-pro-
pagation remaining constant, the inclination of the ripple
will also vary, that is to say, we shall have a *curved* ripple
instead of a straight one. This, of course, is the case which
we find in Nature, but the curvature of such ripples is

totally different from that of the veined structure. Owing
to the quicker translatory movement, the ripples, as they
approach the centre, tend more to parallelism with the
direction of the river; and after having passed the centre,
and reached the slower water near the opposite side, their
inclination to the axis gradually augments. Thus the
ripples from the two sides form a pair of symmetric curves,
which cross each other at the centre, and possess the form
a o b, c o d, shown in Fig. 49. A similar pair of curves

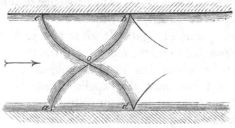

Fig. 49.

would be produced by the reflection of these. Knowing
the variation of motion from side to centre, any competent
mathematician could find the equation of the ripple-curves;
but it would be out of place for me to attempt it here.

THE VEINED STRUCTURE AND PRESSURE.

(30.)

IF a prism of glass be pressed by a sufficient weight, the particles in the line of pressure will be squeezed more closely together, while those at right angles to this line will be forced further apart. The existence of this state of strain may be demonstrated by the action of such squeezed glass upon polarised light. It gives rise to colours, and it is even possible to infer from the tint the precise amount of pressure to which the glass is subjected.   M. Wertheim indeed has most ably applied these facts to the construction of a dynanometer, or instrument for measuring pressures, exceeding in accuracy any hitherto devised.

When the pressure applied becomes too great for the glass to sustain, it flies to pieces. But let us suppose the sides of the prism defended by an extremely strong jacket, in which the prism rests like a closely-fitting plug, and which yields only when a pressure more than sufficient to crush the glass is applied. Let the pressure be gradually augmented until this point is attained; afterwards both the glass and its jacket will shorten and widen; the jacket will yield laterally, being pushed out with extreme slowness by the glass within.

Now I believe that it would be possible to make this experiment in such a manner that the glass should be *flattened*, partly through rupture, and partly through lateral molecular yielding; the prism would change its form, and yet present a firmly coherent mass when removed from its jacket. I have never made the experiment; nobody has, as far as I know; but experiments of this kind are often made by Nature. In the Museum of the Government

School of Mines, for example, we have a collection of quartz stones placed there by Mr. Salter, and which have been subjected to enormous pressure in the neighbourhood of a fault. These rigid pebbles have, in some cases, been squeezed against each other so as to produce mutual flattening and indentation. Some of them have yielded along planes passing through them, as if one-half had slidden over the other; but the reattachment is very strong. Some of the larger stones, moreover, which have endured pressure at a particular point, are fissured radially around this point. In short, the whole collection is a most instructive example of the manner and extent to which one of the most rigid substances in nature can yield on the application of a sufficient force.

Let a prism of ice at 32° be placed in a similar jacket to that which we have supposed to envelop the glass prism. The ice yields to the pressure with incomparably greater ease than the glass ; and if the force be slowly applied, the lateral yielding will far more closely resemble that of a truly plastic body. Supposing such a piece of ice to be filled with numerous small air-bubbles, the tendency of the pressure would be to flatten these bubbles, and to squeeze them out of the ice. Were the substance perfectly homogeneous, this flattening and expulsion would take place uniformly throughout its entire mass; but I believe there is no such homogeneous substance in nature;—the ice will yield at different places, leaving between them spaces which are comparatively unaffected by the pressure. From the former spaces the air-bubbles will be more effectually expelled; and I have no doubt that the result of such pressure acting upon ice so protected would be to produce a laminated structure somewhat similar to that which it produces in those bodies which exhibit slaty cleavage.

I also think it certain that, in this lateral displacement of the particles, these must move past each other.

This is an idea which I have long entertained, as he following passage taken from the paper published by Mr. Huxley and myself will prove :—" Three principal causes may operate in producing cleavage: first, the reducing of surfaces of weak cohesion to parallel planes; second, the flattening of minute cavities; and third, the weakening of cohesion by tangential action. The third action is exemplified by the state of the rails near a station where a break is habitually applied to a locomotive. In this case, while the weight of the train presses vertically, its motion tends to cause longitudinal sliding of the particles of the rail. Tangential action does not, however, necessarily imply a force of the latter kind. When a solid cylinder an inch in height is squeezed to a vertical cake a quarter of an inch in height, it is impossible, physically speaking, that the particles situated in the same vertical line shall move laterally with the same velocity; but if they do not, the cohesion between them will be weakened or ruptured. The pressure, however, will produce new contact ; and if this have a cohesive value equal to that of the old contact, no cleavage from this cause can arise. The relative capacities of different substances for cleavage appears to depend in a great measure upon their different proporties in this respect. In butter, for example, the new attachments are equal, or nearly so, to the old, and the cleavage is consequently indistinct; in wax this does not appear to be the case, and hence may arise in a great degree the perfection of its cleavage. The further examination of this subject promises interesting results." I would dwell upon this point the more distinctly as the advocates of differential motion may deem it to be in their favour; but it appears to me that the mechanical conceptions implied in the above passage are totally different from theirs. If they think otherwise, then it seems to me that they should change the expressions which refer the differential motion to a "drag" towards the centre, and the structure to the sliding of "filaments" past each other in consequence of this drag. Such filamentary sliding may take place in a truly viscous body, but it does not take place in ice.

In one particular the ice resembles the butter referred to in the above quotation; for its new attachments appear to be equal to the old, and this, I think, is to be ascribed to its perfect regelation. As justly pointed out by Mr. John Ball, the veined ice of a glacier, if unweathered, shows no tendency to cleave; for though the expulsion of the air-bubbles has taken place, the reattachment of the particles is so firm as to abolish all evidence of cleavage. When the ice, on the contrary, is weathered, the plates become detached, and I have often been able to split such ice into thin tablets having an area of two or three square feet.

In his Thirteenth Letter Professor Forbes throws out a new and possibly a pregnant thought in connexion with the veins. If I understand him aright—and I confess it is usually a matter of extreme difficulty with me to make sure of this—he there refers the veins, not to the expulsion of the air from the ice, but to its redistribution. The pressure produces "*lines of tearing* in which the air is distributed in the form of regular globules." I do not know what might be made of this idea if it were developed, but at present I do not see how the supposed action could produce the blue bands; and I agree with Professor Wm. Thomson in regarding the explanation as improbable.*

* For an extremely ingenious view of the origin of the veined structure, I would refer to a paper by Professor Thomson, in the ' Proceedings of the Royal Society,' April, 1858.

THE VEINED STRUCTURE AND THE LIQUEFAC-
TION OF ICE BY PRESSURE.

(31.)

I HAVE already noticed an important fact for which we are indebted to Mr. James Thomson, and have referred to the original communications on the subject. I shall here place the physical circumstances connected with this fact before my reader in the manner which I deem most likely to interest him.

When a liquid is heated, the attraction of the molecules operates against the action of the heat, which tends to tear them asunder. At a certain point the force of heat triumphs, the cohesion is overcome, and the liquid boils. But supposing we assist the attraction of the molecules by applying an external pressure, the difficulty of tearing them asunder will be increased; more heat will be required for this purpose; and hence we say that the *boiling point* of the liquid has been *elevated* by the pressure.

If molten sulphur be poured into a bullet-mould, it will be found on cooling to contract, so as to leave a large hollow space in the middle of each sphere. Cast musket-bullets are thus always found to possess a small cavity within them produced by the contraction of the lead. Conceive the bullet placed within its mould and the latter heated; to produce fusion it is necessary that the sulphur or the lead should *swell*. Here, as in the case of the heated water, the tendency to expand is opposed by the attraction of the molecules; with a certain amount of heat however this attraction is overcome and the solid *melts*. But suppose we assist the molecular attraction by a suitable force applied externally, a greater amount of heat than before will be necessary to tear them asunder; and hence

we say that the *fusing point* has been *elevated* by the pressure. This fact has been experimentally established by
Messrs. Hopkins and Fairbairn, who applied to spermaceti
and other substances pressures so great as to raise their
points of fusion a considerable number of degrees.

Let us now consider the case of the metal bismuth. If
the molten metal be poured into a bullet-mould it will
expand on solidifying. I have myself filled a strong cast-
iron bottle with the metal, and found its expansion on
cooling sufficiently great to split the bottle from neck to
bottom. Hence, in order to fuse the bismuth the substance
must *contract;* and it is manifest that an external pressure
which tends to squeeze the molecules more closely together
here *assists* the heat instead of opposing it. Hence, to
fuse bismuth under great pressure, a less amount of
heat will be required than when the pressure is removed ;
or in other words, the fusing point of bismuth is *lowered*
by the pressure. Now, in passing from the solid to the
liquid state, *ice*, like bismuth, contracts, and, if the contraction be promoted by external pressure, as shown by
the Messrs. Thomson, a less amount of heat suffices to
liquefy it.

These remarks will enable us to understand a singular
effect first obtained by myself at the close of 1856 or in
January 1857, noticed at the time in the 'Proceedings' of
the Royal Society, and afterwards fully described in a
paper presented to the Society in December of that year.
A cylinder of clear ice two inches high and an inch in
diameter was placed between two slabs of box-wood, and
subjected to a gradual pressure. I watched the ice in a
direction perpendicular to its length, and saw cloudy lines
drawing themselves across it. As the pressure continued,
these lines augmented in numbers, until finally the prism
presented the appearance of a crystal of gypsum whose
planes of cleavage had been forced out of optical contact.

T

When looked at obliquely it was found that the lines were merely the sections of flat dim surfaces, which lay like

Fig. 50.

Fig. 51.

laminæ one over the other throughout the length of the prism. Fig. 50 represents the prism as it appeared when looked at in a direction perpendicular to its axis; Fig. 51 shows the appearance when viewed obliquely.*

At first sight it might appear as if air had intruded itself between the separated surfaces of the ice, and to test this point I placed a cylinder two inches long and an inch wide upright in a copper vessel which was filled with ice-cold water. The ice cylinder rose about half an inch above the surface of the water. Placing the copper vessel on a slab of wood, and a second slab on the top of the cylinder of ice, the latter was subjected to the gradual action of a small hydraulic press. When the hazy surfaces were well developed in the portion of the ice above the water, the cylinder was removed and examined: the planes of rupture extended throughout the entire length of the cylinder, just as if it had been squeezed in air. I subsequently placed the ice in a stout vessel of glass, and squeezed it, as in the last experiment: the surfaces of discontinuity were seen forming *under the liquid* quite as distinctly as in air.

To prove that the surfaces were due to compression and not to any tearing asunder of the mass by tension, the following experiment was made :—A cylindrical piece of ice, one of whose ends, however, was not parallel to the other, was placed between the slabs of wood, and subjected to

* This effect projected upon a screen is a most striking and instructive class experiment.

pressure. Fig. 52 shows the disposition of the experiment. The effect upon the ice-cylinder was that shown in

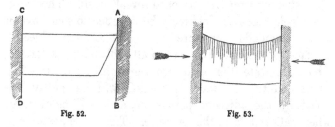

Fig. 52.   Fig. 53.

Fig. 53, the surfaces being developed along that side which had suffered the pressure. On examining the surfaces by a pocket lens they resembled the effect produced upon a smooth cold surface by breathing on it.

The surfaces were always dim ; and had the spaces been filled with air, or were they simply vacuous, the reflection of light from them would have been so copious as to render them much more brilliant than they were observed to be. To examine them more particularly I placed a concave mirror so as to throw the diffused daylight from a window full upon the cylinder. On applying the pressure dim spots were sometimes seen forming in the very middle of the ice, and these as they expanded laterally appeared to be in a state of intense motion, which followed closely the edge of each surface as it advanced through the solid ice. Once or twice I observed the hazy surfaces pioneered through the mass by dim offshoots, apparently liquid, and constituting a kind of decrystallization. From the closest examination to which I was able to subject them, the surfaces appeared to me to be due to internal lique- faction ; indeed, when the melting point of ice, having already a temperature of 32°, is lowered by pressure, its excess of heat must instantly be applied to produce this effect.

I have already given a drawing (p. 386) showing the deve-

T 2

lopment of the veined structure at the base of the ice-cascade of the Rhone; and if we compare that diagram with Fig. 53 a striking similarity at once reveals itself. The ice of the glacier must undoubtedly be liquefied to some extent by the tremendous pressure to which it is here subjected. Surfaces of discontinuity will in all probability be formed, which facilitate the escape of the imprisoned air. The small quantity of water produced will be partly imbibed by the adjacent porous ice, and will be refrozen when relieved from the pressure. This action, associated with that ascribed to pressure in the last section, appears to me to furnish a complete physical explanation of the laminated structure of glacier-ice.

WHITE ICE-SEAMS IN THE GLACIER DU GEANT.

(32.)

On the 28th of July, 1857, while engaged upon the Glacier du Géant, my attention was often attracted by protuberant ridges of what at first appeared to be pure white snow, but which on examination I found to be compact ice filled with innumerable round air-cells; and which, in virtue of its greater power of resistance to wasting, often rose to a height of three or four feet above the general level of the ice. As I stood amongst these ridges, they appeared detached and without order of arrangement, but looked at from a distance they were seen to sweep across the proper Glacier du Géant in a direction concentric with its dirt-bands and its veined structure. In some cases the seams were admirable indications of the relative displacement of two adjacent portions of the glacier, which were divided from each other by a crevasse. Usually the sections of a seam exposed on the opposite sides of a fissure accurately faced each other, and the direction of the seam on both sides was continuous; but at other places they demonstrated the existence of lateral faults, being shifted asunder laterally through spaces varying from a few inches to six or seven feet.

On the following day I was again upon the same glacier, and noticed in many cases the white ice-seams exquisitely honeycombed. The case was illustrative of the great difference between the absorptive power of the ice itself and of the objects which lie upon its surface. Deep cylindrical cells were produced by spots of black dirt which had been scattered upon the surface of the white ice, and which sank to a depth of several inches into the mass. I examined several sections of the veins, and in general I found that their deeper portions blended gradually with the ice on either side of them. But higher up the glacier I found that the veins penetrated only to a limited depth, and did not therefore form an integrant portion of the glacier.

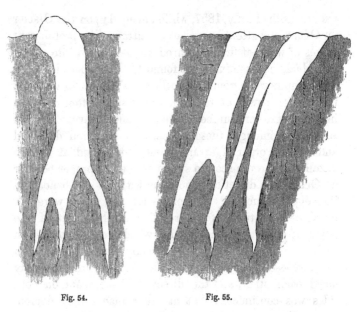

Fig. 54.   Fig. 55.

Figs. 54 and 55 show the sections of two of the seams which were exposed on the wall of a crevasse at some distance below the great ice-fall of the Glacier du Géant.

It was at the base of the Talèfre cascade that the explanation of these curious seams presented itself to me. In one of my earliest visits to this portion of the glacier I

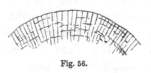

Fig. 56.

was struck by a singular disposition of the blue veins on the vertical wall of a crevasse. Fig. 58 will illustrate what I saw. The veins, within a short distance, dipped *backward* and *forward*, like the junctions of stones used to turn an arch. In some cases I found this variation of the structure so great as to pass in a short distance from the vertical to the horizontal, as shown in Fig. 57.

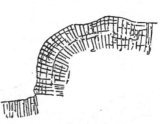

Fig. 57.

Further examination taught me that the glacier here is crumpled in a most singular manner; doubtless by the great pressure to which it is exposed. The following illustration will convey a notion of its aspect: Let one hand be laid flat upon a table, palm downwards, and let the fingers be bent until the space between the first joint and the ends of the fingers is vertical; one of the crumples to which I refer will then be represented. The ice seems bent like the fingers, and the crumples of the glacier are cut by crevasses, which are accurately typified by the spaces between the fingers. Let the second hand now be placed upon the first, as the latter is upon the table, so that the tops of the bent fingers of the second hand shall rest upon the roots of the first: two crumples would thus be formed; a series of such protuberances, with steep fronts, follow each other from the base of the Talèfre cascade for some distance downwards.

On Saturday the 1st of August I ascended these rounded terraces in succession, and observed among them an extremely remarkable disposition of the structure. Fig. 58 is a section of a series of three of the crumples, on which the shading lines represent the direction of the blue veins. At the base of each protuberance I found a seam of white ice wedged firmly into the glacier, and *each of the seams marked a place of dislocation of the veins.* The white seams thinned off gradually, and finally vanished where the violent crumpling of the ice disappeared. In Fig. 59 I

have sketched the wall of a crevasse, which represents
what may be regarded as the incipient crumpling. The

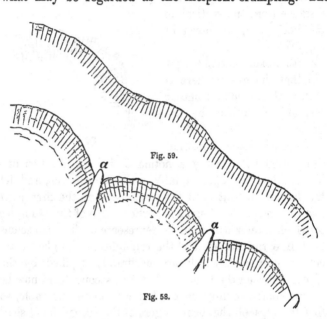

Fig. 59.

Fig. 58.

undulating line shows the contour of the surface, and the
shading lines the veins. It will be observed that the
direction of the veins yields in conformity with the undulation of the surface; and an augmentation of the effect
would evidently result in the crumples shown in Fig. 58.
The appearance of the white seams at those places where
a dislocation occurred was, as far as I could observe, inva-
riable; but in a few instances the seams were observed
upon the platforms of the terraces, and also upon their slopes.
The width of a seam was very irregular, varying from a
few inches at some places to three or four feet at others.

On the 3rd of August I was again at the base of the
Talèfre cascade, and observed a fact, the significance of
which had previously escaped me. The rills which ran

down the ice-slopes collected at the base of each protuberance into a stream, which, at the time of my visit, had
hollowed out for itself a deep channel in the ice. At some
places the stream widened, at others its banks of ice approached each other, and rapids were produced; in fact,
*the channel of such streams appeared to be the exact moulds
of the seams of white ice.*

Instructed thus far, I ascended the Glacier du Géant on
the 5th of August, and then observed on the wrinkles of
this glacier the same leaning backwards and forwards of
the blue veins as I had previously observed upon the
Talèfre. I also noticed on this day that a seam of white
ice would sometimes open out into two branches, which,
after remaining for some distance separate, would reunite
and thus enclose a little glacier-island. At other places
lateral branches were thrown off from the principal seam,
thus suggesting the form of a glacier-rivulet which had
been fed by tributary branches. On the 7th of August
I hunted the seams still farther up the glacier; and found
them at one place descending a steep ice-hill, being crossed
by other similar bands, which however were far less white
and compact. I followed these new bands to their origin,
and found it to be a system of crevasses formed at the
summit of the hill, some of which were filled with snow.
Lower down the crevasses closed, and the snow thus
jammed between their walls was converted into white ice.
These seams, however, never attained the compactness and
prominence of the larger ones which had their origin far
higher up. I singled out one of the best of the latter, and
traced it through all the dislocation and confusion of the
ice, until I found it to terminate in a cavity filled with
snow.

This was near the base of the *séracs*, and the streams
here were abundant. Comparing the shapes of some of
them with that of the ice-bands lower down the glacier, a

T 3

striking resemblánce was observed.　Fig. 60 is the plan of
a deep-cut channel through which a stream flowed on the

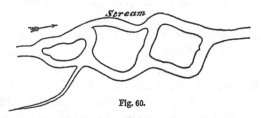

Fig. 60.

day to which I now refer.　Fig. 61 is the plan of a seam of
white ice sketched on the same day, low down upon the

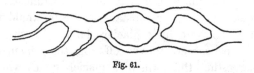

Fig. 61.

glacier.　Instances of this kind might be multiplied; and
the result, I think, renders it certain that the white ice-
seams referred to are due to the filling up of the channels
of glacier-streams by snow during winter, and the subsequent compression of the mass to ice during the descent
of the glacier.　I have found such seams at the bases of all
cascades that I have visited; and in all cases they appear
to be due to the same cause.　The depth to which they
penetrate the glacier must be profound, or the *ablation* of
the ice must be less that what is generally supposed; for
the seams formed so high up on the Glacier du Géant may
be traced low down upon the trunk-stream of the Mer de
Glace.*

These observations on the white ice-seams enable us to
add an important supplement to what has been stated
regarding the origin of the dirt-bands of the Mer de

* The more permanent seams may possibly be due to the filling of
the profound crevasses of the cascade.

Glace   The protuberances at the base of the cascade are due not only to the toning down of the ridges produced by the transverse fracture of the glacier at the summit of the fall, but they undergo modifications by the pressure locally exerted at its base.   The state of things represented in Fig. 57 is plainly due to the partial pushing of one crumple over that next in advance of it.   There seems to be a differential motion of the parts of the glacier in the same longitudinal line; showing that upon the general motion of the glacier smaller local motions are superposed.   The occurrence of the seams upon the faces of the slopes seems also to prove that the pressure is competent, in some cases, to cause the bases of the protuberances to swell, so that what was once the base of a crumple may subsequently form a portion of its slope.   Another interesting fact is also observed where the pressure is violent: the crumples *scale off*, bows of ice being thus formed which usually span the crumples over their most violently compressed portions. I have found this scaling off at the bases of all the cascades which I have visited, and it is plainly due to the pressure exerted at such places upon the ice.

(33.)

Not only at the base of its great cascade, but throughout the greater part of its length, the Glacier du Géant is in a state of longitudinal compression.   The meaning of this term will be readily understood: Let two points, for example, be marked upon the axis of the glacier; if these during its descent were drawn wider apart, it would show that the glacier was in a state of longitudinal strain or tension; if they remained at the same distance apart, it would indicate that neither strain nor pressure was exerted; whereas, if the two points approached each other, which could only be by the quicker motion of the hinder one, the existence of longitudinal compression would be thereby demonstrated.

Taking "Le Petit Balmat" with me, to carry my theodolite, I ascended the Glacier du Géant until I came near the place where it is joined by the Glacier des Périades, and whence I observed a patch of fresh green grass upon the otherwise rocky mountain-side. To this point I climbed, and made it the station for my instrument. Choosing a well-defined object at the opposite side of the glacier, I set, on the 9th of August, in the line between this object and the theodolite, three stakes, one in the centre of the glacier, and the other two at opposite sides of the centre and about 100 yards from it. This done, I descended for a quarter of a mile, when I again climbed the flanking rocks, placing my theodolite in a couloir, down which stones are frequently discharged from the end of a secondary glacier which hangs upon the heights above. Here, as before, I fixed three stakes, chiselled a mark upon the granite, so as to enable me to find the place, and regained the ice without accident. A day or two previously we had set out a third line at some distance lower down, and I was thus furnished with a succession of points along the glacier, the relative motions of which would decide whether it was *pressed* or *stretched* in the direction of its length. On the 10th of August Mr. Huxley joined us; and on the following day we all set out for the Glacier du Géant, to measure the progress of the stakes which I had fixed there. Hirst remained upon the glacier to measure the displacements; I shouldered the theodolite; and Huxley was my guide to the mountain-side, sounding in advance of me the treacherous-looking snow over which we had to pass.

Calling the central stake of the highest line No. 1, that of the middle line No. 2, and that of the line nearest the

Tacul No. 3, the following are the spaces moved over by these three points in twenty-four hours:—

	Inches.	Distances asunder.
No. 1.	20·55	545 yards.
No. 2.	15·43	
No. 3.	12·75	487 yards.

Here we have the fact which the aspect of the glacier suggested. The first stake moves five inches a day more than the second, and the second nearly three inches a day more than the third. As surmised, therefore, the glacier is in a state of longitudinal compression, whereby a portion of it 1000 yards in length is shortened at the rate of eight inches a day.

In accordance with this result, the transverse undulations of the Glacier du Géant, described in the chapter upon Dirt-Bands, *shorten* as they descend. A series of three of them measured along the axis of the glacier on the 6th of August, 1857, gave the following respective lengths:— 955 links, 855 links, 770 links, the shortest undulation being the farthest from the origin of the undulations. This glacier then constitutes a vast ice-press, and enables us to test the explanation which refers the veined structure of the ice to pressure. The glacier itself is transversely laminated, as already stated; and in many cases a structure of extreme definition and beauty is developed in the compressed snow, which constitutes the seams of white ice. In 1857 I discovered a well-developed lenticular structure in some of these seams. In 1858 I again examined them. Clearing away the superficial portions with my axe, I found, drawn through the body of the seams, long lines of blue ice of exquisite definition; in fact, I had never seen the structure so delicately exhibited. The seams, moreover, were developed in portions of the white ice which were near the *centre* of the glacier, and where consequently filamentous sliding was entirely out of the question.

PARTIAL SUMMARY.

1. GLACIERS are derived from mountain snow, which has been consolidated to ice by pressure.

2. That pressure is competent to convert snow into ice has been proved by experiment.

3. The power of yielding to pressure diminishes as the mass becomes more compact; but it does not cease even when the substance has attained the compactness which would entitle it to be called ice.

4. When a sufficient depth of such a substance collects upon the earth's surface, the lower portions are squeezed out by the pressure of the superincumbent mass. If it rests upon a slope it will yield principally in the direction of the slope, and move downwards.

5. In addition to this, the whole mass slides bodily along its inclined bed, and leaves the traces of its sliding on the rocks over which it passes, grinding off their asperities, and marking them with grooves and scratches in the direction of the motion.

6. In this way the deposit of consolidated and unconsolidated snow which covers the higher portions of lofty mountains moves slowly down into an adjacent valley, through which it descends as a true glacier, partly by sliding and partly by the yielding of the mass itself.

7. Several valleys thus filled may unite in a single valley, the tributary glaciers welding themselves together to form a trunk-glacier.

8. Both the main valley and its tributaries are often sinuous, and the tributaries must change their direction to form the trunk; the width of the valley often varies. The glacier is forced through narrow gorges, widening after it has passed them; the centre of the glacier moves more

quickly than the sides, and the surface more quickly than
the bottom; the point of swiftest motion follows the same
law as that observed in the flow of rivers, shifting from
one side of the centre to the other as the flexure of the
valley changes.

9. These various effects may be reproduced by experiments on small masses of ice. The substance may moreover be moulded into vases and statuettes. Straight bars
of it may be bent into rings, or even coiled into knots.

10. Ice capable of being thus moulded is practically
incapable of being stretched. The condition essential to
success is that the particles of the ice operated on shall be
kept in close contact, so that when old attachments have
been severed new ones may be established.

11. The nearer the ice is to its melting point in temperature, the more easily are the above results obtained;
when ice is many degrees below its freezing point it is
crushed by pressure to a white powder, and is not capable
of being moulded as above.

12. Two pieces of ice at 32° Fahr., with moist surfaces,
when placed in contact freeze together to a rigid mass;
this is called Regelation.

13. When the attachments of pressed ice are broken, the
continuity of the mass is restored by the regelation of the
new contiguous surfaces. Regelation also enables two
tributary glaciers to weld themselves to form a continuous
trunk; thus also the crevasses are mended, and the dislocations of the glacier consequent on descending cascades
are repaired. This healing of ruptures extends to the
smallest particles of the mass, and it enables us to account
for the continued compactness of the ice during the descent
of the glacier.

14. The quality of viscosity is practically absent in
glacier-ice. Where pressure comes into play the phenomena are suggestive of viscosity, but where tension comes

into play the analogy with a viscous body breaks down. When subjected to strain the glacier does not yield by stretching, but by breaking; this is the origin of the crevasses.

15. The crevasses are produced by the mechanical strains to which the glacier is subjected. They are divided into marginal, transverse, and longitudinal crevasses; the first produced by the oblique strain consequent on the quicker motion of the centre; the second by the passage of the glacier over the summit of an incline; the third by pressure from behind and resistance in front, which causes the mass to split at right angles to the pressure.

16. The moulins are formed by deep cracks intersecting glacier rivulets. The water in descending such cracks scoops out for itself a shaft, sometimes many feet wide, and some hundreds of feet deep, into which the cataract plunges with a sound like thunder. The supply of water is periodically cut off from the moulins by fresh cracks, in which new moulins are formed.

17. The lateral moraines are formed from the débris which loads the glacier along its edges; the medial moraines are formed on a trunk-glacier by the union of the lateral moraines of its tributaries; the terminal moraines are formed from the débris carried by the glacier to its terminus, and there deposited. The number of medial moraines on a trunk glacier is always one less than the number of tributaries.

18. When ordinary lake-ice is intersected by a strong sunbeam it liquefies so as to form flower-shaped figures within the mass; each flower consists of six petals with a vacuous space at the centre; the flowers are always formed parallel to the planes of freezing, and depend on the crystallization of the substance.

19. Innumerable liquid disks, with vacuous spots, are also formed by the solar beams in glacier ice. These empty

spaces have been hitherto mistaken for air-bubbles, the flat form of the disks being erroneously regarded as the result of pressure.

20. These disks are indicators of the intimate constitution of glacier-ice, and they teach us that it is composed of an aggregate of parts with surfaces of crystallization in all possible planes.

21. There are also innumerable small cells in glacier-ice holding air and water; such cells also occur in lake-ice; and here they are due to the melting of the ice in contact with the bubble of air. Experiments are needed on glacier-ice in reference to this point.

22. At a free surface within or without, ice melts with more ease than in the centre of a compact mass. The motion which we call heat is less controlled at a free surface, and it liberates the molecules from the solid condition sooner than when the atoms are surrounded on all sides by other atoms which impede the molecular motion. Regelation is the complementary effect to the above; for here the superficial portions of a mass of ice are made virtually central by the contact of a second mass.

23. The dirt-bands have their origin in the ice-cascades. The glacier, in passing the brow, is transversely fractured; ridges are formed with hollows between them; these transverse hollows are the principal receptacles of the fine débris scattered over the glacier; and after the ridges have been melted away, the dirt remains in successive stripes upon the glacier.

24. The ice of many glaciers is laminated, and when weathered may be cloven into thin plates. In the sound ice the lamination manifests itself in blue stripes drawn through the general whitish mass of the glacier; these blue veins representing portions of ice from which the air-bubbles have been more completely expelled. This is the veined structure of the ice. It is divided into marginal,

transverse, and longitudinal structure; which may be regarded as complementary to marginal, longitudinal, and transverse crevasses. The latter are produced by tension the former by pressure, which acts in two different ways: firstly, the pressure acts upon the ice as it has acted upon rocks which exhibit the lamination technically called cleavage; secondly, it produces partial liquefaction of the ice. The liquid spaces thus formed help the escape of the air from the glacier; and the water produced, being refrozen when the pressure is relieved, helps to form the blue veins.

APPENDIX.

COMPARATIVE VIEW OF THE CLEAVAGE OF CRYSTALS AND SLATE-ROCKS.

A LECTURE DELIVERED AT THE ROYAL INSTITUTION, ON FRIDAY EVENING THE 6TH OF JUNE, 1856.*

WHEN the student of physical science has to investigate the character of any natural force, his first care must be to purify it from the mixture of other forces, and thus study its simple action. If, for example, he wishes to know how a mass of water would shape itself, supposing it to be at liberty to follow the bent of its own molecular forces, he must see that these forces have free and undisturbed exercise. We might perhaps refer him to the dew-drop for a solution of the question; but here we have to do, not only with the action of the molecules of the liquid upon each other, but also with the action of gravity upon the mass, which pulls the drop downwards and elongates it. If he would examine the problem in its purity, he must do as Plateau has done, withdraw the liquid mass from the action of gravity, and he would then find the shape of the mass to be perfectly spherical. Natural processes come to us in a mixed manner, and to the uninstructed mind are a mass of unintelligible confusion. Suppose half-a-dozen of the best musical performers to be placed in the same room, each playing his own instrument to perfection: though each individual instrument might be a well-spring of melody, still the mixture of all would produce mere noise. Thus it is with the processes of nature. In nature, mechanical and molecular laws mingle, and create apparent confusion. Their mixture constitutes what may be called the *noise* of natural laws, and it is the vocation of the man of science to resolve this noise into its components, and thus to detect the "music" in which the foundations of nature are laid.

The necessity of this detachment of one force from all other forces is nowhere more strikingly exhibited than in the phænomena of crystallization. I have here a solution of sulphate of soda. Prolonging the mental vision beyond the boundaries of sense, we see the atoms of that liquid, like squadrons under the eye of an experienced general, arrang-

* Referred to in the Introduction.

ing themselves into battalions, gathering round a central standard, and forming themselves into solid masses, which after a time assume the visible shape of the crystal which I here hold in my hand. I may, like an ignorant meddler wishing to hasten matters, introduce confusion into this order. I do so by plunging this glass rod into the vessel. The consequent action is not the pure expression of the crystalline forces; the atoms rush together with the confusion of an unorganized mob, and not with the steady accuracy of a disciplined host. Here, also, in this mass of bismuth we have an example of this confused crystallization; but in the crucible behind me a slower process is going on : here there is an architect at work " who makes no chips, no din," and who is now building the particles into crystals, similar in shape and structure to those beautiful masses which we see upon the table. By permitting alum to crystallize in this slow way, we obtain these perfect octahedrons; by allowing carbonate of lime to crystallize, nature produces these beautiful rhomboids ; when silica crystallizes, we have formed these hexagonal prisms capped at the ends by pyramids ; by allowing saltpetre to crystallize, we have these prismatic masses; and when carbon crystallizes, we have the diamond. If we wish to obtain a perfect crystal, we must allow the molecular forces free play : if the crystallizing mass be permitted to rest upon a surface it will be flattened, and to prevent this a small crystal must be so suspended as to be surrounded on all sides by the liquid, or, if it rest upon the suface, it must be turned daily so as to present all its faces in succession to the working builder. In this way the scientific man nurses these children of his intellect, watches over them with a care worthy of imitation, keeps all influences away which might possibly invade the strict morality of crystalline laws, and finally sees them developed into forms of symmetry and beauty which richly reward the care bestowed upon them.

In building up crystals, these little atomic bricks often arrange themselves into layers which are perfectly parallel to each other, and which can be separated by mechanical means ; this is called the cleavage of the crystal. I have here a crystallized mass which has thus far escaped the abrading and disintegrating forces which, sooner or later, determine the fate of sugar-candy. If I am skilful enough, I shall discover that this crystal of sugar cleaves with peculiar facility in one direction. Here, again, I have a mass of rock-salt : I lay my knife upon it, and with a blow cleave it in this direction ; but I find on further examining this substance that it cleaves in more directions than one. Laying my knife at right angles to its former position, the crystal cleaves again ; and, finally placing the knife at right angles to the two former positions, the mass cleaves again. Thus rock-salt cleaves in three directions, and the resulting solid is this perfect cube, which may be broken up into any number of smaller cubes. Here is a mass of Iceland spar, which also cleaves in three directions, not at right angles, but obliquely to each other, the resulting solid being a rhomboid. In each of these cases the mass cleaves with equal facility in all three directions. For the sake of completeness, I may say that many substances cleave with unequal facility in different directions, and the heavy spar I hold in my hand presents an example of this kind of cleavage.

Turn we now to the consideration of some other phænomena to which the term cleavage may be applied. This piece of beech-wood cleaves with facility parallel to the fibre, and if our experiments were fine enough we should discover that the cleavage is most perfect when the edge of the axe is laid across the rings which mark the growth of the tree. The fibres of the wood lie side by side, and a comparatively small force is sufficient to separate them. If you look at this mass of hay severed from a rick, you will see a sort of cleavage developed in it also; the stalks lie in parallel planes, and only a small force is required to separate them laterally. But we cannot regard the cleavage of the tree as the same in character as the cleavage of the hayrick. In the one case it is the atoms arranging themselves according to organic laws which produce a cleavable structure; in the other case the easy separation in a certain direction is due to the mechanical arrangement of the coarse sensible masses of the stalks of hay.

In like manner I find that this piece of sandstone cleaves parallel to the planes of bedding. This rock was once a powder, more or less coarse, held in mechanical suspension by water. The powder was composed of two distinct parts, fine grains of sand and small plates of mica. Imagine a wide strand covered by a tide which holds such powder in suspension:* how will it sink? The rounded grains of sand will reach the bottom first, the mica afterwards, and when the tide recedes we have the little plates shining like spangles upon the surface of the sand. Each successive tide brings its charge of mixed powder, deposits its duplex layer day after day, and finally masses of immense thickness are thus piled up, which, by preserving the alternations of sand and mica, tell the tale of their formation. I do not wish you to accept this without proof. Take the sand and mica, mix them together in water, and allow them to subside, they will arrange themselves in the manner I have indicated; and by repeating the process you can actually build up a sandstone mass which shall be the exact counterpart of that presented by nature, as I have done in this glass jar. Now this structure cleaves with readiness along the planes in which the particles of mica are strewn. Here is a mass of such a rock sent to me from Halifax: here are other masses from the quarries of Over Darwen in Lancashire. With a hammer and chisel you see I can cleave them into

* I merely use this as an illustration; the deposition may have really been due to sediment carried down by rivers. But the action must have been periodic, and the powder duplex.

flags; indeed these flags are made use of for roofing purposes in the districts from which the specimens have come, and receive the name of "slate-stone." But you will discern, without a word from me, that this cleavage is not a crystalline cleavage any more than that of a hay-rick is. It is not an arrangement produced by molecular forces; indeed it would be just as reasonable to suppose that on this jar of sand and mica the particles arranged themselves into layers by the forces of crystallization, instead of by the simple force of gravity, as to imagine that such a cleavage as this could be the product of crystallization.

This, so far as I am aware of, has never been imagined, and it has been agreed among geologists not to call such splitting as this cleavage at all, but to restrict the term to a class of phænomena which I shall now proceed to consider.

Those who have visited the slate quarries of Cumberland and North Wales will have witnessed the phænomena to which I refer. We have long drawn our supply of roofing-slates from such quarries; schoolboys ciphered on these slates, they were used for tombstones in churchyards, and for billiard-tables in the metropolis; but not until a comparatively late period did men begin to inquire how their wonderful structure was produced. What is the agency which enables us to split Honister Crag, or the cliffs of Snowdon, into laminæ from crown to base? This question is at the present moment one of the greatest difficulties of geologists, and occupies their attention perhaps more than any other. You may wonder at this. Looking into the quarry of Penrhyn, you may be disposed to explain the question as I heard it explained two years ago. "These planes of cleavage," said a friend who stood beside me on the quarry's edge, "are the planes of stratification which have been lifted by some convulsion into an almost vertical position." But this was a great mistake, and indeed here lies the grand difficulty of the problem. These planes of cleavage stand in most cases at a high angle to the bedding. Thanks to Sir Roderick Murchison, who has kindly permitted me the use of specimens from the Museum of Practical Geology (and here I may be permitted to express my acknowledgments to the distinguished staff of that noble establishment, who, instead of considering me an intruder, have welcomed me as a brother), I am able to place the proof of this before you. Here is a mass of slate in which the planes of bedding are distinctly marked; here are the planes of cleavage, and you see that one of them makes a large angle with the other. The cleavage of slates is therefore not a question of stratification, and the problem which we have now to consider is, "By what cause has this cleavage been produced?"

In an able and elaborate essay on this subject in 1835, Professor Sedgwick proposed the theory that cleavage is produced by the action of crystalline or polar forces after the mass has been consolidated. "We may affirm," he says, "that no retreat of the parts, no contraction of dimensions in passing to a solid state can explain such phænomena. They appear to me only resolvable on the supposition that crystalline or polar forces acted upon the whole mass simultaneously in one direction and with adequate force." And again, in another place: "Crystalline forces have rearranged whole mountain-masses, producing a beautiful crystalline cleavage, passing alike through all the strata." * The utterance of such a man struck deep, as was natural, into the minds of geologists, and at the present day there are few who do not entertain this view either in whole or in part.† The magnificence of the theory, indeed, has in some cases caused speculation to run riot, and we have books published, aye and largely sold, on the action of polar forces and geologic magnetism, which rather astonish those who know something about the subject. According to the theory referred to, miles and miles of the districts of North Wales and Cumberland, comprising huge mountain-masses, are neither more nor less than the parts of a gigantic crystal. These masses of slate were originally fine mud; this mud is composed of the broken and abraded particles of older rocks. It contains silica, alumina, iron, potash, soda, and mica, mixed in sensible masses mechanically together. In the course of ages the mass became consolidated, and the theory before us assumes that afterwards a process of crystallization rearranged the particles and developed in the mass a single plane of crystalline cleavage. With reference to this hypothesis, I will only say that it is a bold stretch of analogies; but still it has done good service: it has drawn attention to the question; right or wrong, a theory thus thoughtfully uttered has its value; it is a dynamic power which operates against intellectual stagnation; and, even by provoking opposition, is eventually of service to the cause of truth. It would, however, have been remarkable, if, among the ranks of geologists themselves, men were not found to seek an explanation of the phænomena in question, which involved a less hardy spring on the part of the speculative faculty than the view to which I have just referred.

The first step in an inquiry of this kind is to put oneself into contact with nature, to seek facts. This has been done, and the labours of Sharpe (the late President of the Geological Society, who, to the loss of science and the sorrow of all who knew him, has so suddenly been

---

* 'Transactions of the Geological Society,' Ser. ii. vol. iii. p. 477.

† In a letter to Sir Charles Lyell, dated from the Cape of Good Hope, February 20, 1836, Sir John Herschel writes as follows:—"If rocks have been so heated as to allow of a commencement of crystallization, that is to say, if they have been heated to a point at which the particles can begin to move amongst themselves or at least on their own axes, some general law must then determine the position in which these particles will rest on cooling. Probably that position will have some relation to the direction in which the heat escapes. Now when all or a majority of particles of the same nature have a general tendency to one position, that must of course determine a cleavage plane."

taken away from us), Sorby, and others, have furnished us with a body
of evidence which reveals to us certain important physical phænomena,
associated with the appearance of slaty cleavage, if they have not produced it. The nature of this evidence we will now proceed to consider.

Fossil shells are found in these slate-rocks. I have here several
specimens of such shells, occupying various positions with regard to the
cleavage planes. They are squeezed, distorted, and crushed. In some
cases a flattening of the convex shell occurs, in others the valves are
pressed by a force which acted in the plane of their junction, but in all
cases the distortion is such as leads to the inference that the rock
which contains these shells has been subjected to enormous pressure
in a direction at right angles to the planes of cleavage; the shells
are all flattened and spread out upon these planes. I hold in my
hand a fossil trilobite of normal proportions. Here is a series of fossils
of the same creature which have suffered distortion. Some have
lain across, some along, and some oblique to the cleavage of the slate
in which they are found; in all cases the nature of the distortion
is such as required for its production a compressing force acting at right
angles to the planes of cleavage. As the creatures lay in the mud in
the manner indicated, the jaws of a gigantic vice appear to have closed
upon them and squeezed them into the shape you see. As further
evidence of the exertion of pressure, let me introduce to your notice a
case of contortion which has been adduced by Mr. Sorby. The bedding of the rock shown in this figure* was once horizontal; at A we
have a deep layer of mud, and at $m\,n$ a layer of comparatively unyielding gritty material; below that again, at B, we have another layer
of the fine mud of which slates are formed. This mass cleaves along
the shading lines of the diagram; but look at the shape of the intermediate bed: it is contorted into a serpentine form, and leads irre-
sistibly to the conclusion that the mass has been pressed together at
right angles to the planes of cleavage. This action can be experimentally imitated, and I have here a piece of clay in which this is done
and the same result produced on a small scale. The amount of compression, indeed, might be roughly estimated by supposing this contorted bed $m\,n$ to be stretched out, its length measured and compared
with the distance $c\,d$; we find in this way that the yielding of the
mass has been considerable.

Let me now direct your attention to another proof of pressure. You
see the varying colours which indicate the bedding on this mass of
slate. The dark portion, as I have stated, is gritty, and composed of
comparatively coarse particles, which, owing to their size, shape, and
gravity, sink first and constitute the bottom of each layer. Gradually
from bottom to top the coarseness diminishes, and near the upper
surface of each layer we have a mass of comparatively fine clean mud.

* Omitted here.

Sometimes this fine mud forms distinct layers in a mass of slate-rock, and it is the mud thus consolidated from which are derived the German razor-stones, so much prized for the sharpening of surgical instruments. I have here an example of such a stone. When a bed is thin, the clean white mud is permitted to rest, as in this case, upon a slab of the coarser slate in contact with it: when the bed is thick, it is cut into slices which are cemented to pieces of ordinary slate, and thus rendered stronger. The mud thus deposited sometimes in layers is, as might be expected, often rolled up into nodular masses, carried forward; and de-posited by the rivers from which the slate-mud has subsided. Here, indeed, are such nodules enclosed in sandstone. Everybody who has ciphered upon a school-slate must remember the whitish-green spots which sometimes dotted the surface of the slate; he will remember how his slate-pencil usually slid over such spots as if they were greasy. Now these spots are composed of the finer mud, and they could not, on account of their fineness, *bite* the pencil like the surrounding gritty portions of the slate. Here is a beautiful example of the spots: you observe them on the cleavage surface in broad patches: but if this mass has been compressed at right angles to the planes of cleavage, ought we to expect the same marks when we look at the edge of the slab? The nodules will be flattened by such pressure, and we ought to see evidence of this flattening when we turn the slate edgeways. Here it is. The section of a nodule is a sharp ellipse with its major axis parallel to the cleavage. There are other examples of the same nature on the table; I have made excursions to the quarries of Wales and Cumberland, and to many of the slate-yards of London, but the same fact invariably appears, and thus we elevate a common experience of our boyhood into evidence of the highest significance as regards one of the most important problems of geology. In examining the magnetism of these slates, I was led to infer that these spots would contain a less amount of iron than the surrounding dark slate. The analysis was made for me by Mr. Hambly in the laboratory of Dr. Percy at the School of Mines. The result which is stated in this Table justifies the conclusion to which I have referred.

Analysis of Slate.

Purple Slate. Two Analyses.

1. Percentage of iron	..	..	.. 5·85
2. „ „	..	..	.. 6·13
Mean	..	..	.. 5·99

Greenish Slate.

1. Percentage of iron	..	..	.. 3·24
2. „ „	..	..	.. 3·12
Mean	..	..	.. 3·18

U

The quantity of iron in the dark slate immediately adjacent to the greenish spot is, according to these analyses, nearly double of the quantity contained in the spot itself. This is about the proportion which the magnetic experiments suggested.

Let me now remind you that the facts which I have brought before you are typical facts—each is the representative of a class. We have seen shells crushed; the unhappy trilobites squeezed, beds contorted, nodules of greenish marl flattened; and all these sources of independent testimony point to one and the same conclusion, namely, that slate-rocks have been subjected to enormous pressure in a direction at right angles to the planes of cleavage.*

In reference to Mr. Sorby's contorted bed, I have said that by supposing it to be stretched out and its length measured, it would give us an idea of the amount of yielding of the mass above and below the bed. Such a measurement, however, would not quite give the amount of yielding; and here I would beg your attention to a point, the significance of which has, so far as I am aware of, hitherto escaped attention. I hold in my hand a specimen of slate, with its bedding marked upon it; the lower portions of each bed are composed of a comparatively coarse gritty material, something like what you may suppose this contorted bed to be composed of. Well, I find that the cleavage takes a bend in crossing these gritty portions, and that the tendency of these portions is to cleave more at right angles to the bedding. Look to this diagram: when the forces commenced to act, this intermediate bed, which though comparatively unyielding is not entirely so, suffered longitudinal pressure; as it bent, the pressure became gradually more lateral, and the direction of its cleavage is exactly such as you would infer from a force of this kind—it is neither quite across the bed, nor yet in the same direction as the cleavage of the slate above and below it, but intermediate between both. Supposing the cleavage to be at right angles to the pressure, this is the direction which it ought to take across these more unyielding strata.

Thus we have established the concurrence of the phænomena of cleavage and pressure—that they accompany each other; but the question still remains, Is this pressure of itself sufficient to account for the cleavage? A single geologist, as far as I am aware, answers boldly in the affirmative. This geologist is Sorby, who has attacked the question in the true spirit of a physical investigator. You remember the cleavage of the flags of Halifax and Over Darwen, which is caused by the interposition of plates of mica between the layers. Mr. Sorby examines the structure of slate-rock, and finds plates of mica to be a constituent.

* While to my mind the evidence in proof of pressure seems perfectly irresistible, I by no means assert that the manner in which I stated it is incapable of modification. All that I deem important is the fact that pressure has been exerted; and provided this remain firm, the fate of any minor portion of the evidence by which it is here established is of comparatively little moment.

He asks himself, what will be the effect of pressure upon a mass containing such plates confusedly mixed up in it? It will be, he argues— and he argues rightly—to place the plates with their flat surfaces more or less perpendicular to the direction in which the pressure is exerted. He takes scales of the oxide of iron, mixes them with a fine powder, and, on squeezing the mass, finds that the tendency of the scales is to set themselves at right angles to the line of pressure. Now the planes in which these plates arrange themselves will, he contends, be those along which the mass cleaves.

I could show you, by tests of a totally different character from those applied by Mr. Sorby, how true his conclusion is, that the effect of pressure on elongated particles or plates will be such as he describes it. Nevertheless, while knowing this fact, and admiring the ability with which Mr. Sorby has treated this question, I cannot accept his explanation of slate-cleavage. I believe that even if these plates of mica were wholly absent, the cleavage of slate-rocks would be much the same as it is at present.

I will not dwell here upon minor facts,—I will not urge that the perfection of the cleavage bears no relation to the quantity of mica present; but I will come at once to a case which to my mind completely upsets the notion that such plates are a necessary element in the production of cleavage.

Here is a mass of pure white wax: there are no mica particles here; there are no scales of iron, or anything analogous mixed up with the mass. Here is the self-same substance submitted to pressure. I would invite the attention of the eminent geologists whom I see before me to the structure of this mass. No slate ever exhibited so clean a cleavage; it splits into laminæ of surpassing tenuity, and proves at a single stroke that pressure is sufficient to produce cleavage, and that this cleavage is independent of the intermixed plates of mica assumed in Mr. Sorby's theory. I have purposely mixed this wax with elongated particles, and am unable to say at the present moment that the cleavage is sensibly affected by their presence,—if anything, I should say they rather impair its fineness and clearness than promote it.

The finer the slate the more perfect will be the resemblance of its cleavage to that of the wax. Compare the surface of the wax with the surface of this slate from Borrodale in Cumberland. You have precisely the same features in both: you see flakes clinging to the surfaces of each, which have been partially torn away by the cleavage of the mass: I entertain the conviction that if any close observer compares these two effects, he will be led to the conclusion that they are the product of a common cause.*

* I have usually softened the wax by warming it, kneaded it with the fingers, and pressed it between thick plates of glass previously wetted. At the ordinary

But you will ask, how, according to my view, does pressure produce
this remarkable result? This may be stated in a very few words.

Nature is everywhere imperfect! The eye is not perfectly achro-
matic, the colours of the rose and tulip are not pure colours, and the
freshest air of our hills has a bit of poison in it. In like manner there
is no such thing in nature as a body of perfectly homogeneous structure.
I break this clay which seems so intimately mixed, and find that the
fracture presents to my eyes innumerable surfaces along which it has
given way, and it has yielded along these surfaces because in them the
cohesion of the mass is less than elsewhere. I break this marble, and
even this wax, and observe the same result: look at the mud at the
bottom of a dried pond; look to some of the ungravelled walks in
Kensington Gardens on drying after rain,—they are cracked and split,
and other circumstances being equal, they crack and split where the
cohesion of the mass is least. Take then a mass of partially consolidated mud. Assuredly such a mass is divided and subdivided by
surfaces along which the cohesion is comparatively small. Penetrate
the mass, and you will see it composed of numberless irregular nodules
bounded by surfaces of weak cohesion. Figure to your mind's eye
such a mass subjected to pressure,—the mass yields and spreads out in
the direction of least resistance; * the little nodules become converted
into laminæ, separated from each other by surfaces of weak cohesion,
and the infallible result will be that such a mass will cleave at right
angles to the line in which the pressure is exerted.

Further, a mass of dried mud is full of cavities and fissures. If you
break dried pipe-clay you see them in great numbers, and there are
multitudes of them so small that you cannot see them. I have here a
piece of glass in which a bubble was enclosed; by the compression of
the glass the bubble is flattened, and the sides of the bubble approach
each other so closely as to exhibit the colours of thin plates. A similar
flattening of the cavities must take place in squeezed mud, and this
must materially facilitate the cleavage of the mass in the direction
already indicated.

Although the time at my disposal has not permitted me to develop
this thought as far as I could wish, yet for the last twelve months the
subject has presented itself to me almost daily under one aspect or
another. I have never eaten a biscuit during this period in which an
intellectual joy has not been superadded to the more sensual pleasure,
for I have remarked in all such cases cleavage developed in the mass by

summer-temperature the wax is soft, and tears rather than cleaves; on this
account I cool my compressed specimens in a mixture of pounded ice and salt, and
when thus cooled they split beautifully.

* It is scarcely necessary to say that if the mass were squeezed equally in *all*
directions no laminated structure could be produced; it must have room to yield
in a lateral direction.

the rolling-pin of the pastrycook or confectioner. I have only to break these cakes, and to look at the fracture, to see the laminated structure of the mass; nay, I have the means of pushing the analogy further: I have here some slate which was subjected to a high temperature during the conflagration of Mr. Scott Russell's premises. I invite you to compare this structure with that of a biscuit; air or vapour within the mass has caused it to swell, and the mechanical structure it reveals is precisely that of a biscuit. I have gone a little into the mysteries of baking while conducting my inquiries on this subject, and have received much instruction from a lady-friend in the manufacture of puff-paste. Here is some paste baked in this house under my own superintendence. The cleavage of our hills is accidental cleavage, but this is cleavage with intention. The volition of the pastrycook has entered into the forma-tion of the mass, and it has been his aim to preserve a series of surfaces of structural weakness, along which the dough divides into layers. Puff-paste must not be handled too much, for then the continuity of the surfaces is broken; it ought to be rolled on a cold slab, to prevent the butter from melting and diffusing itself through the mass, thus rendering it more homogeneous and less liable to split. This is the whole philosophy of puff-paste; it is a grossly exaggerated case of slaty cleavage.

As time passed on, cases multiplied, illustrating the influence of pressure in producing lamination. Mr. Warren De la Rue informs me that he once wished to obtain white-lead in a fine granular state, and to accomplish this he first compressed the mass: the mould was conical, and permitted the mass to spread a little laterally under the pressure. The lamination was as perfect as that of slate, and quite defeated him in his effort to obtain a granular powder. Mr. Brodie, as you are aware, has recently discovered a new kind of graphite: here is the substance in powder, of exquisite fineness. This powder has the pecu-liarity of clinging together in little confederacies; it cannot be shaken asunder like lycopodium; and when the mass is squeezed, these groups of particles flatten, and a perfect cleavage is produced. Mr. Brodie himself has been kind enough to furnish me with specimens for this evening's lecture. I will cleave them before you: you see they split up into plates which are perpendicular to the line in which the pressure was exerted. This testimony is all the more valuable, as the facts were obtained without any reference whatever to the question of cleavage.

I have here a mass of that singular substance Boghead Cannel. This was once a mass of mud, more or less resembling this one, which I have obtained from a bog in Lancashire. I feel some hesitation in bringing this substance before you, for, as in other cases, so in regard to Boghead Cannel, science—not science, let me not libel it, but the quibbling, litigious, money-loving portion of human nature speaking through the mask of science—has so contrived to split hairs as to render the qualities of the substance somewhat mythical. I shall therefore content myself with showing you how it cleaves, and with expressing my conviction that pressure had a great share in the production of this cleavage.

The principle which I have enunciated is so simple as to be almost trivial; nevertheless, it embraces not only the cases I have mentioned, but, if time permitted, I think I could show you that it takes a much wider range. When iron is taken from the puddling furnace, it is a more or less spongy mass: it is at a welding heat, and at this temperature is submitted to the process of rolling: bright smooth bars such as this are the result of this rolling. But I have said that the mass is more or less spongy or nodular, and, notwithstanding the high heat, these nodules do not perfectly incorporate with their neighbours: what then? You would say that the process of rolling must draw the nodules into fibres—it does so; and here is a mass acted upon by dilute sulphuric acid, which exhibits in a striking manner this fibrous structure. The experiment was made by my friend Dr. Percy, without any reference to the question of cleavage.

Here are other cases of fibrous iron. This fibrous structure is the result of mechanical treatment. Break a mass of ordinary iron and you have a granular fracture; beat the mass, you elongate these granules, and finally render the mass fibrous. Here are pieces of rails along which the wheels of locomotives have slided; the granules have yielded and become plates; they exfoliate or come off in leaves. All these effects belong, I believe, to the great class of phænomena of which slaty cleavage forms the most prominent example.*

Thus, ladies and gentlemen, we have reached the termination of our task. I commenced by exhibiting to you some of the phænomena of crystallization. I have placed before you the facts which are found to be associated with the cleavage of slate-rocks. These facts, as finely expressed by Helmboltz, are so many telescopes to our spiritual vision, by which we can see backward through the night of antiquity, and discern the forces which have been in operation upon the earth's surface

> " Ere the lion roared,
> Or the eagle soared."

From evidence of the most independent and trustworthy character, we come to the conclusion that these slaty masses have been subjected to enormous pressure, and by the sure method of experiment we have

---

* An eminent authority informs me that he believes these surfaces of weak cohesion to be due to the interposition of films of graphite, and not to any tendency of the iron itself to become fibrous: this of course does not in any way militate against the theory which I have ventured to propose. All that the theory requires is surfaces of weak cohesion, however produced, and a change of shape of such surfaces consequent on pressure or rolling.

shown—and this is the only really new point which has been brought before you—how the pressure is sufficient to produce the cleavage. Expanding our field of view, we find the self-same law, whose footsteps we trace amid the crags of Wales and Cumberland, stretching its ubiquitous fingers into the domain of the pastrycook and ironfounder; nay, a wheel cannot roll over the half-dried mud of our streets without revealing to us more or less of the features of this law. I would say, in conclusion, that the spirit in which this problem has been attacked by geologists indicates the dawning of a new day for their science. The great intellects who have laboured at geology, and who have raised it to its present pitch of grandeur, were compelled to deal with the subject in mass; they had no time to look after details. But the desire for more exact knowledge is increasing; facts are flowing in, which, while they leave untouched the intrinsic wonders of geology, are gradually supplanting by solid truths the uncertain speculations which beset the subject in its infancy. Geologists now aim to imitate, as far as possible, the conditions of nature, and to produce her results; they are approaching more and more to the domain of physics; and I trust the day will soon come when we shall interlace our friendly arms across the common boundary of our sciences, and pursue our respective tasks in a spirit of mutual helpfulness, encouragement, and good-will.

INDEX.

———◇———

LONDON . PRINTED BY WILLIAM CLOWES AND SONS, STAMFORD STREET,
AND CHARING CROSS.

Printed in the United States
By Bookmasters